Air Conditioning for Building Engineers and Managers

Air Conditioning for Building Engineers and Managers

—Operation and Maintenance

by Seymour G. Price

Industrial Press Inc., 200 Madison Ave., New York, N.Y. 10016

Air Conditioning for Building Engineers and Managers — Operation and Maintenance $16.50

Library of Congress Catalog Card Number: 79-108194
SBN 8311-3001-6

Contents

Preface

This handbook was prepared for building engineers, superintendents, and property managers concerned with the operation and maintenance of air conditioning equipment in large apartment and commercial buildings, and in public institutions.

The attempt has been made to present the material in an informative, readily understood manner and to avoid, wherever possible, technical explanations beyond the scope of the average building engineer. Yet it offers all of the fundamentals needed to understand the complexities involved in providing human comfort. It will also prove to be a valuable tool in the intelligent day-to-day operation and maintenance of air conditioning equipment with resultant dollar savings.

In the preparation of this handbook, the author has been able to draw on information and material furnished by reliable manufacturers, trade publications, the Department of the Navy, and the Post Office Department which, through his long association with them, has made certain material available for which he wishes to express his appreciation.

The author also wishes to express his appreciation to his fellow workers in the field, to his longtime associate, Albert F. Johnson, for his practical and always helpful advice, and to William M. Schieb, Maintenance Consultant, for his valuable suggestions concerning the scope of the material covered in this handbook.

Introduction

In organizing the material in this handbook, the author has proceeded with the idea of first introducing the reader to the subject of refrigeration in a basic, easily understood manner. From here, the subject of self-contained units, familiar to practically everyone, is covered. With this beginning, central refrigeration plants are introduced, together with their related components, such as water-cooling towers and the water treatment required by them. The next subject covered is air handlers and the special controls serving them. Troubleshooting guides to self-contained units, central refrigeration plants, and water-cooling towers have also been included in the subject matter. Thus, the first six chapters of this handbook should provide building management personnel with a good working knowledge of the operation of air conditioning equipment.

The day-to-day, routine preventive maintenance of equipment is explained in Chapter 7. Sample Maintenance Check Lists have been provided here for an average, electrically-driven refrigeration plant and system of approximately 100 tons, from which check lists for large plants can be developed.

Chapter 8 covers the economical operation of central, *electrically-driven* refrigeration plants. This type of plant is the only one fully described, since it is the type most widely used at present, and is the largest contributor to increasing plant utility bills.

Sample specifications for refrigeration service and water-treatment contracts to assure effective maintenance performance by the contractor, are presented in Chapter 9.

Both employee and management benefit by adhering to safety rules. Some lists of important safety measures with which building management should become familiar are included in Chapter 10.

Chapter 11 has been set up as a complete program for the controlled operation of heating and cooling plants at large commercial buildings and installations. A "refrigeration plant operating log," specifically developed for the program, will supply much of the data otherwise available only by the use of complex control centers costing many thousands of dollars, plus their requiring expensive maintenance. Where such centers have already been installed, the program will still provide important support data. In Chapter 12, Exhibit 12F illustrates the large dollar savings potentially made possible through the use of this program.

CHAPTER 1

Theory of Refrigeration

Basic Theory of Refrigeration

A brief review of the theory of refrigeration is given in this chapter as an aid in properly diagnosing poor performance and failures in equipment. An understanding of the changes occurring in a refrigerant as it passes through the refrigeration cycle will give a better appreciation of the function of each unit in the system.

The transfer of heat between substances and gases is basic to refrigeration. This exchange is controlled by conduction, convection, or radiation induced mechanically or chemically. It is not within the scope of this book to discuss the complex thermodynamics involved; following, however, are a few basic facts derived from scientific conclusions, and stated in simple terms:

a. *Transfer of Heat* — Heat always flows from a warmer to a colder substance or space, and is transferred by conduction, convection, or radiation.

b. *Temperature Changes* — When a substance is either absorbing heat or releasing it, the change in heat content will either produce a change in temperature or a change in the physical state of a substance, as in the case of water being converted to steam or ice. Such changes in state do not necessarily take place at a single fixed temperature. Water will boil at 212 deg. F, provided that the atmospheric pressure is normal; that is, 14.7 pounds per square inch (psi). However, if the water is under a pressure of 67 psi, for example, it must be heated to 300 deg. F before it begins to boil.

c. *Pressure-Temperature Relationship* — From the preceding paragraph, it can be seen that there is a direct relationship between the boiling point of a given liquid and the pressure exerted upon it. Raising the pressure on a liquid raises its boiling point and, conversely — lowering the pressure lowers the boiling point.

d. *Evaporation of Boiling Liquids* — The heat required to change a liquid to vapor is in addition to the heat already applied to bring the liquid to its boiling point. During this change in state, the temperature of the liquid remains constant, even though there is a continued application of heat. Also, the temperature of the vapor remains the same as the temperature of the boiling liquid. It may seem that this extra heat is wasted; however, the extra heat or the energy it represents is being used in the process of converting the liquid into vapor. In other words, during this physical change (from a liquid to a vapor) heat is absorbed without raising the temperature of either the liquid or the vapor. It can be assumed, therefore, that the evaporation of a liquid causes heat to be taken away or withdrawn from any surrounding medium or substance. The science of refrigeration is based on this unique property of the evaporation process and on the fact that liquids can be vaporized at any desired temperature simply by changing the pressure exerted upon them.

e. *Evaporation of Refrigerants* — Because evaporation is necessary if a liquid is to absorb more heat without raising its overall temperature, it might appear that merely boiling the water will serve as a primitive form of refrigeration. This is impractical for a number of reasons, the most obvious being that considerable latent heat of vaporization (energy used to convert water into vapor) has to be absorbed from a surrounding medium that has a temperature above the boiling point of the water. In this case, enough heat to produce this change in state cannot be absorbed from the surrounding air and must be obtained from a high-temperature heat source such as a gas flame, an electric heating element, or a coal fire.

This does not, however, invalidate the principle of heat absorption by evaporation. Ammonia, methyl chloride, carbon dioxide, sulfur dioxide, and the refrigerants in the halogen family boil at temperatures very much lower than that of water. Refrigerant R-12, for example, boils at 21.6 deg. F below zero at normal atmospheric pressure. To prevent these refrigerants from boiling and evaporating away at normal surrounding temperatures, they must be contained under pressure (See b and c, above). When the refrigerants are released from this pressure, as in the evaporator coils of a refrigeration unit, they immediately begin to boil and vaporize. The air or water surrounding the evaporator coils transfers its heat to the refrigerant in a continuing process of vaporizing the liquid; or, to consider the exchange of heat in another way, the refrigerant is constantly drawing heat from the surrounding air or water and absorbing it at a rate equal to the heat needed to maintain the evaporation process. For a transfer of heat to occur, the boiling temperature of the refrigerant, at the pressure in the evaporator, must be lower than the temperature of the surrounding air or water. Thus, we return to the basic principle that heat flows from warmer to cooler substances.

Refrigeration Cycle

Elementary Cycle — Figure 1-1 illustrates an elementary refrigeration cycle in which the refrigerant is allowed to evaporate through boiling, in chamber *A*. The heat necessary for vaporization is taken from the surrounding air, and the vaporized liquid passes through tube *B*, to chamber *C*. A stream of water running over the outside of chamber *C* absorbs the heat taken on by the refrigerant, allowing the refrigerant to condense and again become a liquid.

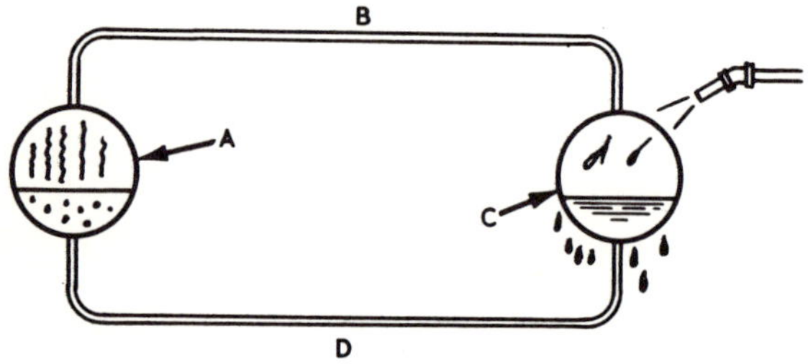

Courtesy of The Trane Company

Fig. 1-1. Elementary refrigeration cycle showing use of latent heat.

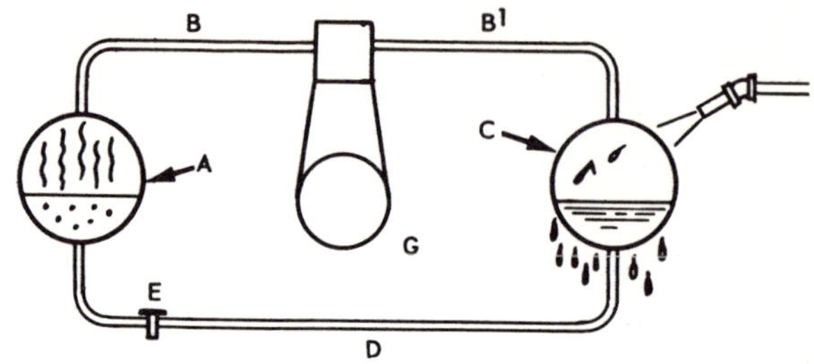

Courtesy of The Trane Company

Fig. 1-2. Addition of compressor to elementary refrigeration cycle.

The water used for condensing must be as cold as the liquid refrigerant in chamber *A*, because the boiling point of a liquid and the condensing temperature of the vapor are exactly the same at any one pressure. The use of the cold water in chamber *A*, therefore, would be just as effective as the refrigerant. It is evident that some method must be applied which will raise the condensing temperature of the refrigerant above the normal temperature of available water or air.

Application of Pressure — The boiling temperature of a liquid can be changed through a wide range by the application of pressure. For the moment, consider that the primary concern in making a simple refrigeration cycle workable is to raise the pressure in chamber *C* to the point where the condensing temperature of the refrigerant entering it will be greater than the temperature of the water sprayed over the outside of the container. In this way, heat can flow from the refrigerant to the water. This, of course, involves the introduction into the system of a pressure-producing device, or compressor. Figure 1-2 illustrates this addition. It also indicates the need of an expansion valve *E* to maintain the higher pressure built up by the action of the compressor *G*. In this system, the evaporated refrigerant is drawn from chamber *A*, placed under pressure by the compressor, and delivered to chamber *C* with its condensing tempera-

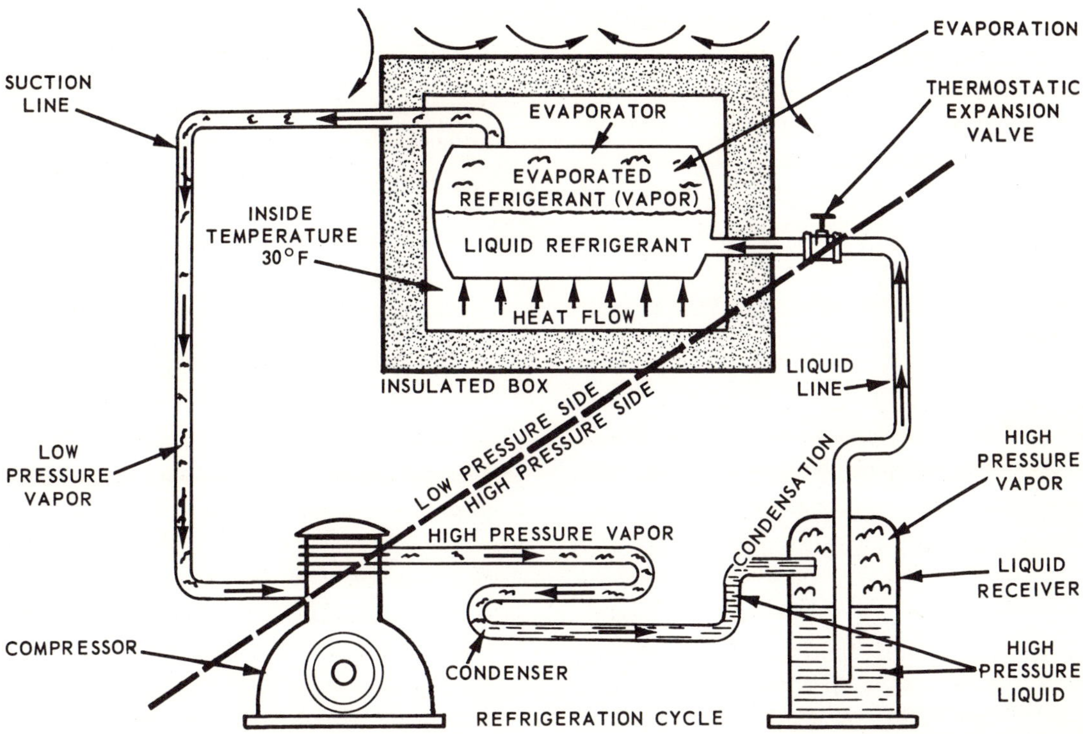

Courtesy of The Trane Company

Fig. 1-3. Low-pressure and high-pressure sections of a simple refrigeration cycle.

ture raised high enough so that water or air, of relatively warm temperature, can condense the refrigerant to a liquid. The refrigerant then flows along line *D* to be metered by the expansion valve into the evaporation chamber, and thus, the cycle begins again.

High- and Low-pressure Separation — The refrigeration cycle or system is equally divided with respect to pressure; one half is a low-pressure section and the other half a high-pressure section. From the compressor through the condenser and back to the expansion valve, constitutes the high-pressure section. These divisions are commonly referred to as the "low side" and the "high side." Figure 1-3 shows a simple schematic drawing of a refrigeration system, illustrating this division.

CHAPTER 2

Self-Contained Units

Definition

A self-contained air conditioning unit consists of a factory-made assembly which normally includes an evaporator or cooling coil, and a compressor and condenser combination. (It may or may not include a heating coil.) To differentiate it from a central installation (referred to in Chapter 3), it may be described as a self-contained unit wherein: (1) cold air is produced by the unit, (2) the air is circulated by the unit, itself, to the area to be cooled, or (3) the air is blown through ducts attached directly to the unit. These units are designed to serve one room or any other enclosed space, and normally range in capacity from ⅓ to 30 tons of refrigeration. They are installed in windows, through the wall, free-standing, or portable. They have DX (direct expansion) coils over which the air is blown.

Window or Through-the-Wall Units

These room coolers usually range from 5,000 to 34,000 Btu (approximately ⅓ to 3 tons). The use of windows for the installation of these units is not an absolute necessity. (See Fig. 2-1.) Since there must be access to outside air, however, for ventilation and exhaust purposes as well as for the air-cooled condenser, they may be installed in transoms or directly in outside walls (commonly called a "through-the-wall" installation). In construction and operating principles, the room cooler is a small and simplified version of certain parts of a more complex central system. As shown in Fig. 2-2 and Fig. 2-3, the basic refrigeration components are all present. Cooling of the condenser coils is accomplished by air. Circulation of the room air is accomplished by a fan that blows across the evaporator coils. Moisture that is con-

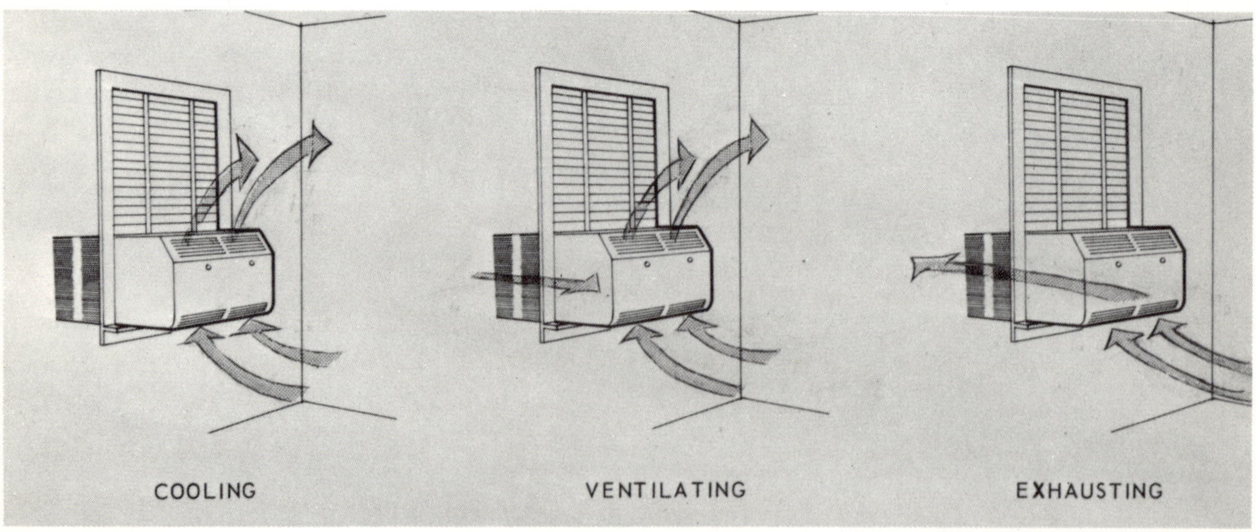

Courtesy of the Department of the Navy

Fig. 2-1. Package type, room air-conditioner showing cooling, ventilating, and exhausting air-flow patterns.

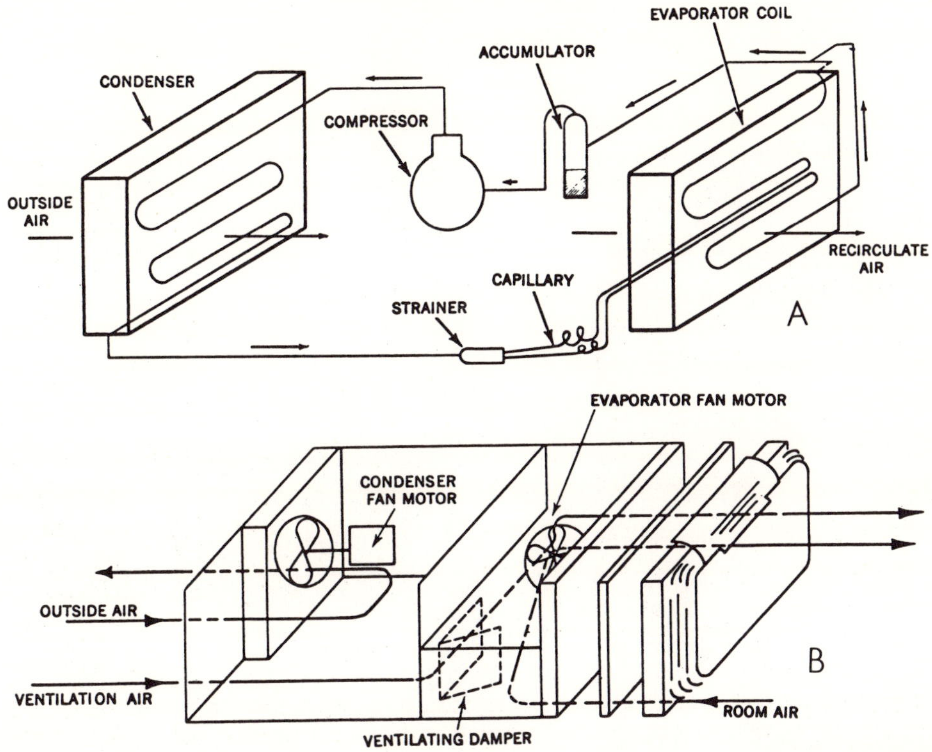

Courtesy of the Department of the Navy

Fig. 2-2. (A) Refrigerant cycle of a room air-conditioner. (B) Air-handling components of a room air-conditioner.

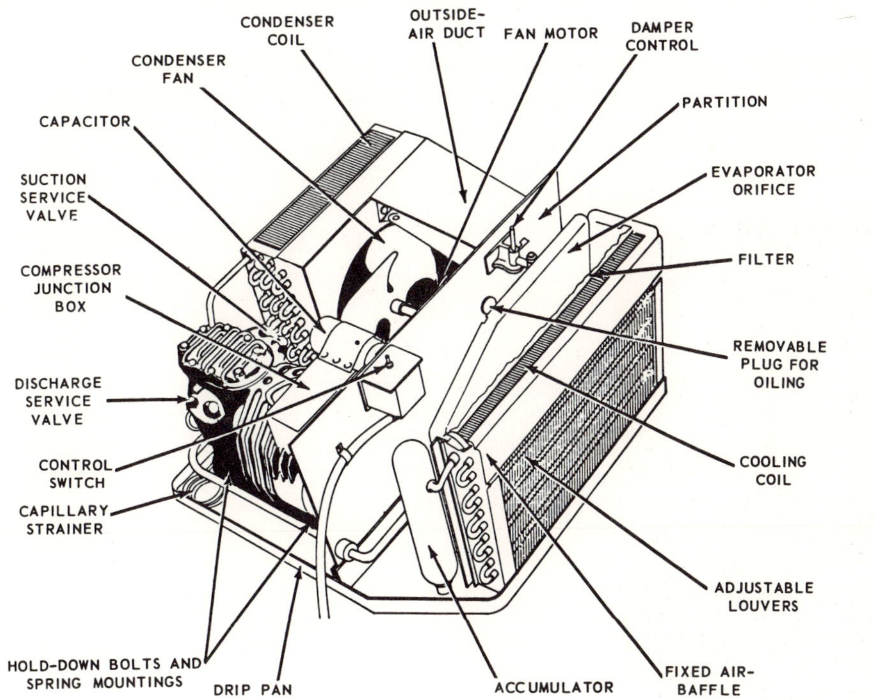

Courtesy of the Department of the Navy

Fig. 2-3. Window-mounted air-conditioner with cabinet removed to show maintenance points.

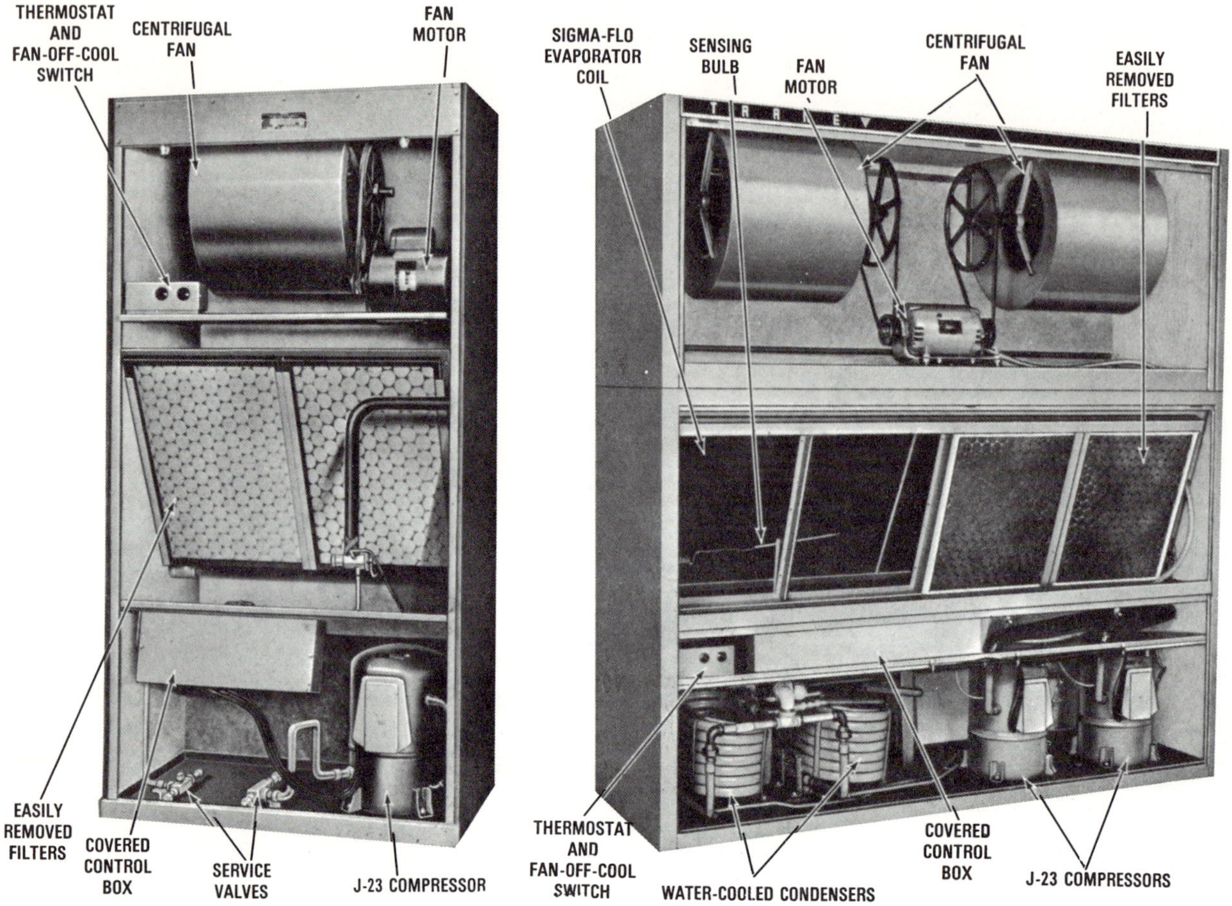

Courtesy of The Trane Company

Fig. 2-4. Typical floor (package) units — open, to show components. (Left) Unit with air-cooled condenser. (Right) Water-cooled condenser. Discharge in these units can be free or ducted from the top or back. Return air can be either to the front or back.

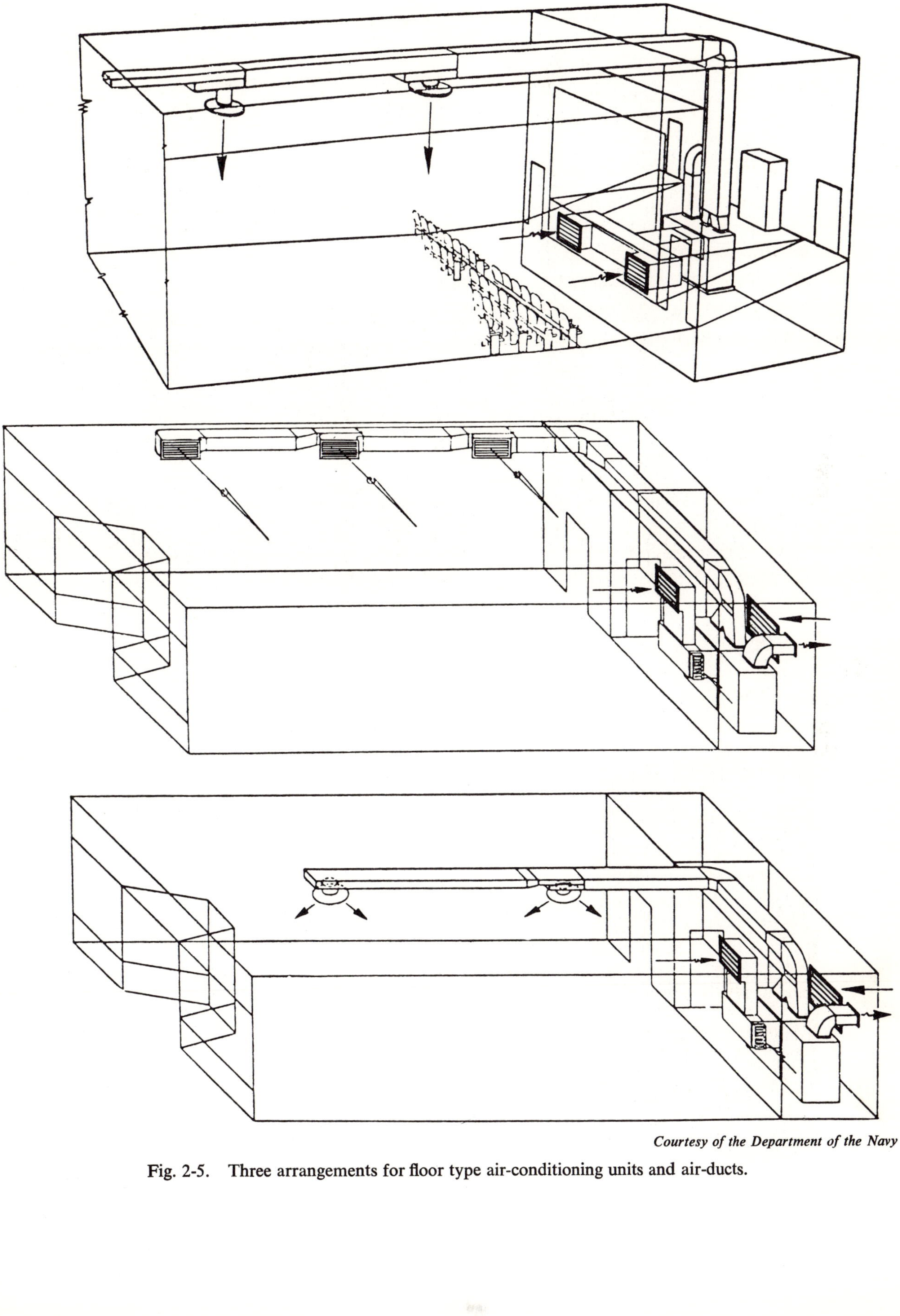

Courtesy of the Department of the Navy

Fig. 2-5. Three arrangements for floor type air-conditioning units and air-ducts.

densed from the humid air by these coils is collected in a pan at the bottom of the unit. It is usually drained toward the back of the unit, where the condenser fan picks up the drainage in the form of droplets and mist, and adds them to the air blast that cools the condenser. Most room coolers are equipped with thermostats that maintain a fixed dry-bulb temperature and moisture content in an area within reasonable limits.

Floor Units

These units normally range in size from 3 to 15 tons, and are sometimes referred to as "package" units. (See Fig. 2-4.) They may be air-cooled or water-cooled. Figure 2-4 shows typical floor units (with the panels removed and major components revealed), which may have either air-cooled condensers, evaporative condensers, or small, water-cooling towers. These may be used

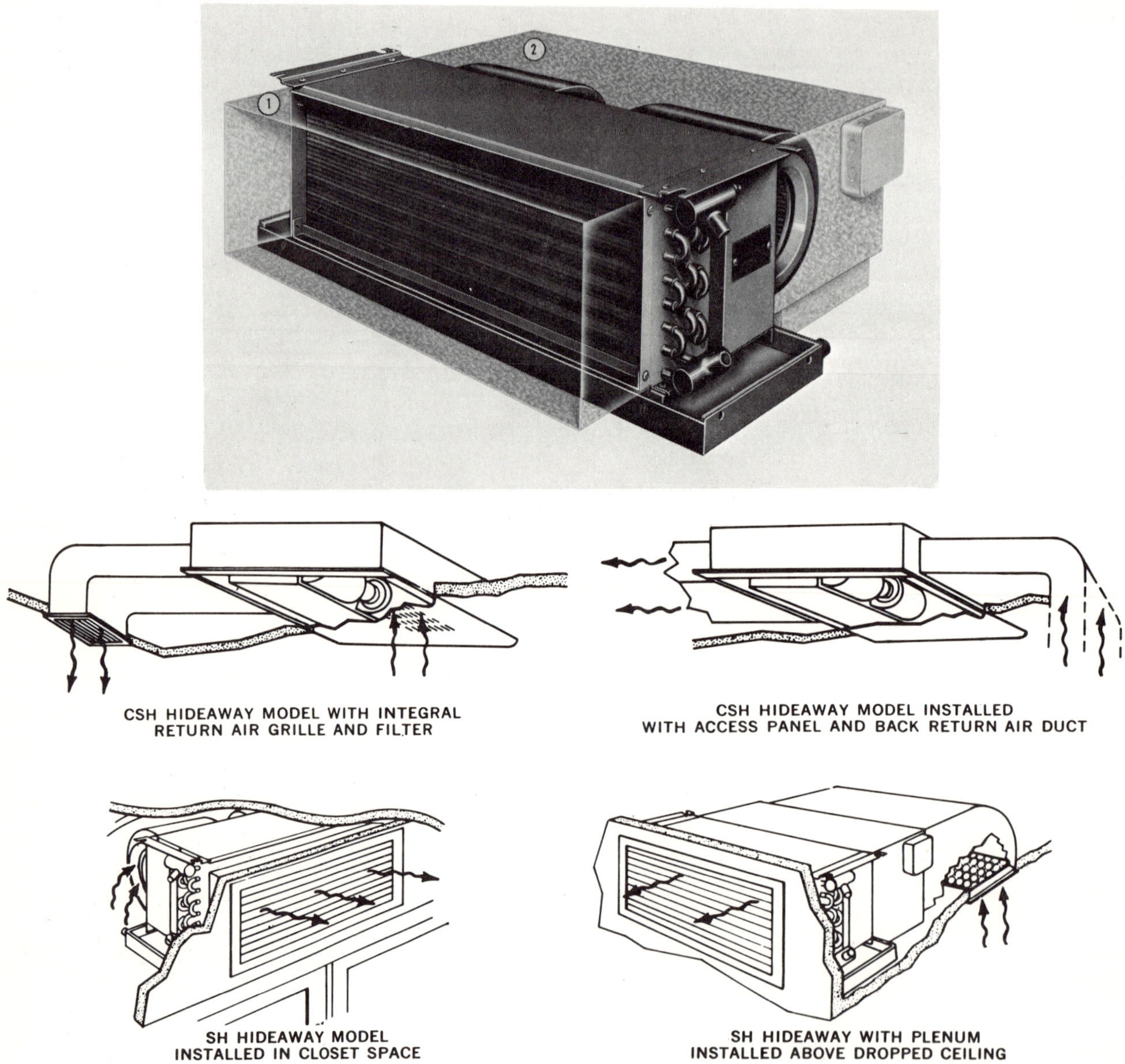

Courtesy of McQuay, Inc.

Fig. 2-6. Adaptability of ceiling-hung air-conditioning units to various conditions. (Top) Optional accessories: (1) 6″ Duct extension, (2) Return-air plenum.

with "free throw," as shown in Fig. 2-4, or with duct work as shown in Fig. 2-5.

Ceiling Units

Normally ranging from ½ to 3 tons, these units are particularly advantageous where floor space is limited. They also adapt themselves to recessed application. (See Fig. 2-6.)

Roof-mounted Units — The range in these units has become greater in the last few years; normally now, from 3 to 15 tons. Shown in Fig. 2-7 is an air-cooled unit which can heat, cool, and ventilate. It is considered self-contained since it meets the definition set forth at the beginning of this chapter.

Capacity Ratings

Until recently, the size or capacity of a self-contained air conditioner was given in terms of (a) tons, or (b) the amount of horsepower developed by its compressor motor. The term "ton" did not apply to the weight of the unit but to a measure of its cooling effect. For example, a ton of refrigeration was considered to be equivalent to the cooling effect that resulted from a ton of ice, melting in 24 hours. Gradually, the terms "horsepower" and "ton" were mistakenly used interchangeably as if they had the same meaning. As a result, a 1-horsepower unit was spoken of as a 1-ton unit, but the terms had little relationship. To establish a uniform method of measuring capacity, most manufacturers have now set up standards for rating and testing their units in terms of British thermal units per hour (Btu/hr.), which is a standard measure of heat. By this method, various makes and models of air conditioners are rated according to the number of Btu of heat they will *remove* from a room in a given period of time. Therefore, when selecting a self-contained unit, it should be made certain that one is getting the most Btu per watt input, based on the unit's cost.

Proper Capacity — Care must be exercised in choosing a room air-conditioner. It will be inefficient and more costly to run if it is too large, yet if too small, it will not do a satisfactory job. The reason for this is very ably explained in the following quote from a publication of the National Better Business Bureau:

> There is, in addition, a close relationship between cooling capacity and dehumidifying action. Suppose, for example, that to serve your particular zone you require a 7,000 Btu/hr. conditioner. Now suppose you install one rated at 10,000 Btu/hr. The large capacity unit would quickly pull the temperature down and the thermostat — which is responsive to heat, not humidity — would automatically shut off both the cooling function and, with it, the dehumidifying action. The room would be cool enough, but it would still be excessively humid. To do the same cooling job, the smaller-capacity unit would operate more of the time, and, therefore, would perform its dehumidifying function properly.
>
> This explains why you may feel "cold and clammy" in an improperly air-conditioned room. If the conditioner is oversized it will cool the air without remaining in operation long enough to dehumidify it.

Wiring Capacity — Room air-conditioners normally operate on 115-, 208-, or 230-volt alternating current. In most localities, it is permissible to connect a unit, designed to operate on 115 volts and 7.5 amperes, directly into the usual 15-ampere circuit. (Some localities require a separate circuit.) It must be made certain that any lamps and appliances already on the circuit, or to be placed on the circuit later, will not exceed a total of 860 watts.

Units operating on 208 or 230 volts must be connected to wiring providing this type of service. The use of Fusetrons or Fustats in air conditioner circuits may be advantageous since they permit an initial surge of current without blowing.

> NOTE: The Association of Home Appliance Manufacturers, Chicago, Ill., issues directories of practically all room air-conditioners on the market. The directories list the name of the manufacturer and certify to the cooling capacity and electric input of each individual unit.

General Maintenance — Operating Checks

Adjustments for Comfort — Those responsible for maintenance can prolong the life of the unit

Courtesy of McQuay, Inc.

Fig. 2-7. Roof-mounted heating and cooling unit which can be used for heating, cooling, or ventilation. (Shown with housing, and with one side removed.)

by making routine checks on its operation. For example, a window-mounted unit is not adjusted properly when cold blasts of air are felt. Better circulation will result if the adjustable louvers of the unit are turned so that the cold air is blown upward. After rising to the ceiling, the air will flow gently down toward the opposite side of the room and no streams of uncomfortable cold air will flow directly at the occupants of the room. If more than the usual number of people is expected to be present, the room should be cooled ahead of time.

Fan Speeds — The conditioner probably has more than one fan speed. On high speed the temperature of the room can be lowered rapidly, but a high noise factor is often produced. Few people realize that more moisture can be removed from the air when the fan is operated at low speed. The more slowly the air moves over the coils, the more moisture is taken from them. Therefore, on a day when the humidity is high, people will be more comfortable sooner, if the fan is operating at low speed.

Excessive Moisture — On sticky, high-humidity days, activities that throw an excessive amount of moisture into the room should be avoided. Washing the floor, hanging wet fabrics indoors, taking prolonged showers, and leaving boiling water uncovered will cause the unit to work harder to overcome these extra sources of humidity.

Waste of Conditioned Air — Maintenance personnel should be on the lookout for any waste of conditioned air. Prevail on occupants of the room to use the Venetian blinds, or whenever possible, shield the unit with an awning. Watch for large gaps under doors that lead to areas which are not air-conditioned. Encourage people to leave air conditioner operating when they go to lunch or when they leave the room for extended periods. Conditioned air will be lost when the system is shut down, and also when the system is turned on again; the compressor will have to work continuously for a long time to bring the temperature and humidity back down. Do not allow tinkering with controls or thermostats by unauthorized persons. (For additional do's and don'ts to assure efficient and economical operation, see last part of Chapter 8.)

General Maintenance

Air-handling Components — The air-handling parts, such as fans, motors, and air filters, and accessories such as dampers, louvers, and grilles require the most attention. A noisy fan, a broken belt, a defective motor, or a dirty filter can cause considerable trouble. (See Fig. 2-3.)

Fans — Keep fan blades clean and do not allow foreign material to build up on them. In cleaning, do not use abrasives that will wear surfaces. If solids form on blades, scrape them off with a putty knife, then clean the blades with detergent and a sponge. Do not bend, twist, or warp blades, and do not allow blades to catch or bind in the housing or scroll. Check washers for wear and replace unserviceable ones. Excessive vibration can damage the bearing shaft on which the fan is mounted. Vibration may be caused by any of the following:

a. Poorly aligned bearings or shaft
b. Bent shaft
c. Unbalance in the fan motor
d. Loose anchor bolts
e. Loose setscrews on the rotor hub
f. Loose bolts on bearings
g. Fan slipping on the motor shaft.

Motors — Check the mounting to see that connections are tight. This does not apply to motors with vibration isolators such as are used with open-type compressor motors and with hermetically sealed compressor-motor combinations. The bolted connections to the vibration isolators of these motors must be loose enough so that the vibration will be absorbed. (See Fig. 2-3.) Keep motors clean and lubricate them according to the manufacturer's instructions. Check for loose electrical connections, switches, and terminals; replace if defective. Check the voltage supply; if it is above or below the rated voltage by as much as 10 per cent, an electrician should correct the voltage supply.

Filters

Checking — Check filters as frequently as necessary to prevent dust and lint from collecting. Clogged filters may cause icing of the coil with resultant damage to the compressor.

Sectional Throw-aways — Remove, discard, and replace. They may be vacuum cleaned *only* in an emergency until a replacement is available.

Sectional Cleanables — Wash filters in a solution of mild detergent and water. A supply of spare filters should be kept on hand so that the air conditioners need not be shut down while the filters are being cleaned. For installations with a large number of filters of this type, a contract for cleaning and servicing them should be considered.

> CAUTION: Under no circumstances shall air conditioners be operated without filters.

Dampers, Louvers, and Grilles

Make periodic checks to determine the need for tightening or replacing broken parts, or repainting surfaces. Keep damper controls in operable condition, and do not allow dirt to accumulate.

Precautionary Advice

In the maintenance of these units, observe the following important points:

a. Do not tamper with hermetically sealed refrigeration equipment, factory-calibrated automatic temperature controls, and other devices covered by warrantees.

b. Never mix different refrigerants in the same system. Operating characteristics of different refrigerants are not similar.

c. The refrigerant should not be allowed to become contaminated with moisture. The refrigerant system should not be filled until the piping is evacuated, cleaned, and dried free of moisture. Use moisture-indicating devices and refrigerant driers of the silica-gel or similar type.

d. *Low Refrigerant Supply* — Maintenance personnel must warn occupants not to operate an air conditioner when the refrigerant supply in the system is very low. One visible indication of this condition is an evaporator covered with frost, an insulator which greatly reduces the cooling capacity of the unit. This could result in motor and compressor breakdown due to non-stop operation of the unit.

e. *Water Treatment* — Required for those units having a water-cooling tower. (See Chapter 5.)

Troubleshooting Guide

Table 2-1, a good troubleshooting guide, begins with page 13. This guide is to be used mainly to help solve problems involving non-operation or faulty operation of a unit. The extent to which it will be used will depend upon the size of the installation, and the know-how of the maintenance personnel. In those cases where the "remedy" calls for the repair of a certain part, the cost between performance by maintenance personnel versus contract service must be evaluated so that the lowest cost is assured. This is also true between repair of certain parts such as motors or compressors, versus replacement.

Life Expectancy (Self-contained units)

Window, and Through-the-wall units	3 to 7 Years
Ceiling units	4 to 8 Years
Floor units (Small)	6 to 12 Years
Roof-mounted units	9 to 13 Years
Package units (Large)	10 to 15 Years

Table 2-1. Troubleshooting Guide for Self-Contained Air Conditioners

Type of unit	Complaint	Cause	Possible remedy
With open-type compressor	Electric motor will not start	Power failure	Check circuit for power source
		Compressor stuck	Locate cause and repair
		Belt too tight	Adjust belt tension
		Manual reset in starter open	Determine cause of overload and repair. Reset overload cutout
		Thermostat setting too high	Lower thermostat setting
		Low voltage	Check with voltmeter, then call power company
		Burned-out motor	Repair or replace
		Frozen compressor caused by locked or damaged mechanism	Remove and repair compressor
	Unit cycles on and off	Intermittent power interruption	Tighten connections or replace defective power supply parts
		High-pressure cutout defective	Replace high-pressure cutout
		High-pressure cutout set too low. Overload opens after having been reset	Raise cutout pressure. Check voltage and current drawn
		Leaky liquid-line solenoid valve	Repair or replace
		Dirty or iced evaporator	Clean or defrost evaporator. Check filters and fan drive.
		Overcharge of refrigerant or noncondensible gas	Remove excess refrigerant or purge noncondensible gas
		Lack of refrigerant	Repair refrigerant leak and re-charge
		Restricted liquid-line strainer	Clean strainer
		Faulty motor	Repair or replace faulty motor
	Coil frosts	Filters dirty	Clean filters
		Not enough air over coil	Clean or remove restriction from supply or return ducts or grilles
		Defective expansion valve	Replace valve
	Unit runs but will not cool	Unit not fully charged	Recharge slightly, then check for leaks in the refrigerant circuit, then fully charge
		Leaky suction valve or discharge valve	Remove compressor cylinder head and clean or replace valve plate
		Expansion valve not set correctly	Adjust expansion valve
		Strainer clogged	Remove, clean, and replace strainer
		Air in refrigerant circuit. Moisture in expansion-valve orifice	Purge unit of air. Clean orifice and install silica gel dryer
		Flash gas in liquid line	Add refrigerant

Courtesy of the Department of the Navy

Table 2-1 (Continued). Troubleshooting Guide for Self-Contained Air Conditioners

Type of unit	Complaint	Cause	Possible remedy
With open-type compressor (continued)	No air blows from supply grille	Ice or dirt on evaporator	Clean coil or defrost
		Blower belt broken or loose	Adjust belt tension, or replace belt
		Blower bearing frozen	Repair or replace bearing and lubricate as directed
	Discharge pressure too high	Improper operation of condenser	Correct air flow. Clean coil surface
		Air in system	Purge
		Overcharge of refrigerant	Remove excess or purge
	Discharge pressure too low	Lack of refrigerant	Repair leak and charge
		Broken or leaky compressor discharge valves	Remove head, examine valves and replace those found to be operating improperly
	Suction pressure too high	Overfeeding of expansion valve	Regulate superheat setting expansion valve and check to see that remote bulb is properly attached to suction line
		Expansion valve stuck in open position	Repair or replace valve
		Broken suction valves in compressor	Remove head, examine valves and replace those found to be inoperative
	Suction pressure too low	Lack of refrigerant	Repair leak and charge
		Clogged liquid line strainer	Clean strainer
		Expansion valve power assembly has lost charge	Replace expansion-valve power assembly
		Obstructed expansion valve	Clean valve and replace if necessary
		Contacts on control thermostat stuck on closed position	Repair thermostat or replace if necessary
With hermetic motor-compressor combination	Compressor runs continuously; good refrigeration effect	Air over condenser restricted	Remove restriction or provide for more air circulation over the condenser
	Compressor runs continuously; unit is too cold	Thermostatic switch contacts badly burned	Replace thermostatic switch
		Thermostatic switch bulb has become loose	Secure bulb in place
		Thermostatic switch improperly adjusted	Readjust thermostatic switch
	Compressor runs continuously; little refrigeration effect	Extremely dirty condenser	Clean condenser
		No air circulating over condenser	Provide air circulation
		Ambient temperature too high	Provide ventilation or move to a cooler location
		Load too great	Analyze load

Courtesy of the Department of the Navy

Table 2-1 (Continued). Troubleshooting Guide for Self-Contained Air Conditioners

Type of unit	Complaint	Cause	Possible remedy
With hermetic motor-compressor combination (continued)	Compressor runs continuously; no refigeration	A restriction that prevents the refrigerant from entering the evaporator. A restriction is usually indicated by a slight refrigeration effect at the point of restriction	Locate the possible points of restriction, and try jarring with a plastic hammer, or heating to a temperature of about 110 degrees F. If the restriction does not open, replace the unit
		Compressor not pumping. This would be indicated by a cool discharge line and a hot compressor housing. The wattage is generally low.	Replace the unit
		Short of refrigerant	See manufacturer's instructions
	Compressor short cycles, poor refrigeration effect	Loose electrical connections	Locate loose connections and make them secure
		Defective thermostatic switch	Replace thermostatic switch
		Defective motor starter	Replace defective motor starter or relay
		Air restriction at evaporator	Remove air restriction
	Compressor short cycles, no refrigeration	Dirty condenser	Clean the condenser
		Ambient temperature too high	Provide ventilation or move to a cooler location
		Defective wiring	Repair or replace defective wiring
		Thermostatic switch operating erratically	Replace thermostatic switch
		Relay erratic	Replace relay
	Compressor runs too frequently	Poor air circulation around the condenser or too high ambient temperature	Increase the air circulation around the condenser. In some localities the temperature is extremely high, and nothing can be done to correct this
		Load too great. Worn compressor. Generally accompanied by rattles or knocks	Analyze end use. Replace unit or bring it to the shop for repairs
	Compressor does not run	Motor is not operating	If the trouble is outside the sealed unit, it should be corrected; for example, wires should be repaired or replaced and thermostatic switches or relays should be replaced. If the trouble is inside the sealed unit, the sealed unit should be replaced
	Compressor will not run (Assume that the thermostatic switch and relay, and the electric wiring and current supply are in good condition and operating normally)	If the cabinet has been moved, some oil may be on top of the piston	Wait an hour or so, and then attempt to start the motor by turning the current on and off many times. On some compressors, it may be necessary to wait 6 or 8 hours.
		Compressor may be stuck, or some parts may be broken	Replace the unit

Courtesy of the Department of the Navy

Table 2-1 (Concluded). Troubleshooting Guide for Self-Contained Air Conditioners

Type of unit	Complaint	Cause	Possible remedy
With hermetic motor-compressor combination (continued)	Compressor will not run (Assume that the thermostatic switch and relay, and the electric wiring and current supply are in good condition and operating normally) (continued)	Connections may be broken on the inside of the unit, or the motor winding may be open	Replace the unit. Sometimes after sealed units have been standing idle for a long time the piston may stick in the cylinder wall. It is sometimes possible to start the compressor by turning on the current and bumping the outer housing with a rubber mallet.
	Compressor is unusually hot.	Condenser is dirty, or there is a lack of air circulation	Clean the condenser; increase the air circulation
		Unusually heavy service or load	If possible, decrease load. Perhaps another unit is required
		Low voltage	This could be caused by too small feed wires. If the wires feeding the refrigerating unit become warm, it is an indication that they are too small and should be replaced with larger wires
		A shortage of oil	Add oil if possible; if this is not possible, the unit must be replaced. A shortage of refrigerant will cause a shortage of oil in the crankcase of the compressor
	No refrigeration after starting up after a long shutdown or on delivery	Generally, during a long shutdown, an amount of liquid refrigerant will get into the crankcase of the compressor. When this happens, the compressor operation will cause no noticeable refrigeration effect until all the liquid refrigerant has evaporated from the crankcase	Allow the compressor to operate until its internal heat drives the liquid refrigerant from the crankcase. Under some conditions, this may take as long as 24 hours. This time can be shortened by turning an electric heater on the compressor and raising the compressor temperature, not exceeding 110 degrees F.
	Compressor is noisy	Mountings have become worn or deteriorated. The walls against which the unit is placed may be of an extremely hard surface and may resound and amplify the slight noise from the compressor into the room	Replace the rubber mountings. Place a piece of sound-absorbing material on the wall against which the unit is placed, or move the unit to a new location
		Shortage of oil and/or refrigerant	Add oil and refrigerant if possible. If it is impossible, the unit must be replaced
		The sealed unit mechanism has become worn	Replace the unit
	After each defrosting there is a long on cycle before refrigeration is again normal	Slight shortage of refrigerant	Add refrigerant if possible; if not, replace the unit.
		Condenser is dirty	Clean the condenser
		Thermostatic switch bulb is loose	Secure the bulb in place
		There is a restriction between the receiver or condenser and/or the evaporator	Attempt to remove the restriction by jarring with a plastic hammer or by heating the possible points of restriction to about 110 degrees F. If this does not correct the trouble, the unit must be replaced or brought to the shop for repairs.

Courtesy of the Department of the Navy

CHAPTER 3

Central Chilled-Water Refrigeration Plants

(Serving Chilled Water to Central Air-Handlers and Room Fan-Coil Units)

Definition

The refrigeration unit in this type of installation produces chilled water which is piped to cooling coils in one or more "air handlers" or "room fan-coil units" (See Chapter 6), and is located at some distance from the refrigeration unit. The main difference between a self-contained or packaged air-conditioner and a central installation (as defined herein) is that in the former, *cold air* is produced at the machine whereas in the latter, *chilled water* is produced at the machine.

Accomplishments — With a central installation, temperature, humidity, cleanliness, and distribution are controlled to meet the requirements that are desired for the conditioned space. The air to be conditioned is: (1) taken from outside the building, (2) recirculated from the conditioned space, or (3) supplied from both of these sources. In all instances the air is moved across the cooling coils by the action of the fans. As air passes over the cooling coils, its temperature is lowered and a portion of the moisture contained therein is condensed, thereby lowering the humidity.

Types of Central Plants

Only two types of plant will be covered in this Handbook; one is the medium-to-large plant utilizing a water-cooling tower. The other is a medium-size plant using an air-cooled condenser in lieu of the water-cooling tower. Further, since the great majority of installations at this time are electricity driven, only electrically-driven equipment will be covered in detail. Prime movers other than electric are covered briefly, at the end of this chapter.

Electrically-driven Plant — The components of the medium- to large-size central plants only, will be covered in detail. This will save repetitive explanations since the basic components of the air-cooled plant perform similar functions. Figures 3-1, 3-2, 3-3, 3-4, and 3-5 are illustrations of medium-to-large refrigeration plants utilizing water-cooling towers.

Basic Components

The basic components for the medium-to-large central plant consist of the following:

1. Refrigeration machine (Compressor)
2. Water-cooling tower and condensing-water system
3. Chilled-water system
4. Air-handling system.

These basic components together with related equipment, such as pumps for the chilled- and condensing-water systems, and fan coils, make up that part of an air conditioning system pertaining to cooling and ventilation.

Compressor — The compressor raises the pressure of the refrigerant gas received from the chiller (or evaporator) and delivers it to the condenser at a pressure and temperature at which it is readily condensed to a liquid.

Types of Compressors

Reciprocating — The reciprocating-type compressor utilizes cylinders, pistons, and connecting rods. (See Figs. 3-1, 3-2, and 3-3.) These figures also show the difference between the two main types of reciprocating compressors, "open" and "hermetic." Note that the open type has the compressor driven by the motor through an open crankshaft whereas in the hermetic type the motor and compressor are in the same housing and the motor

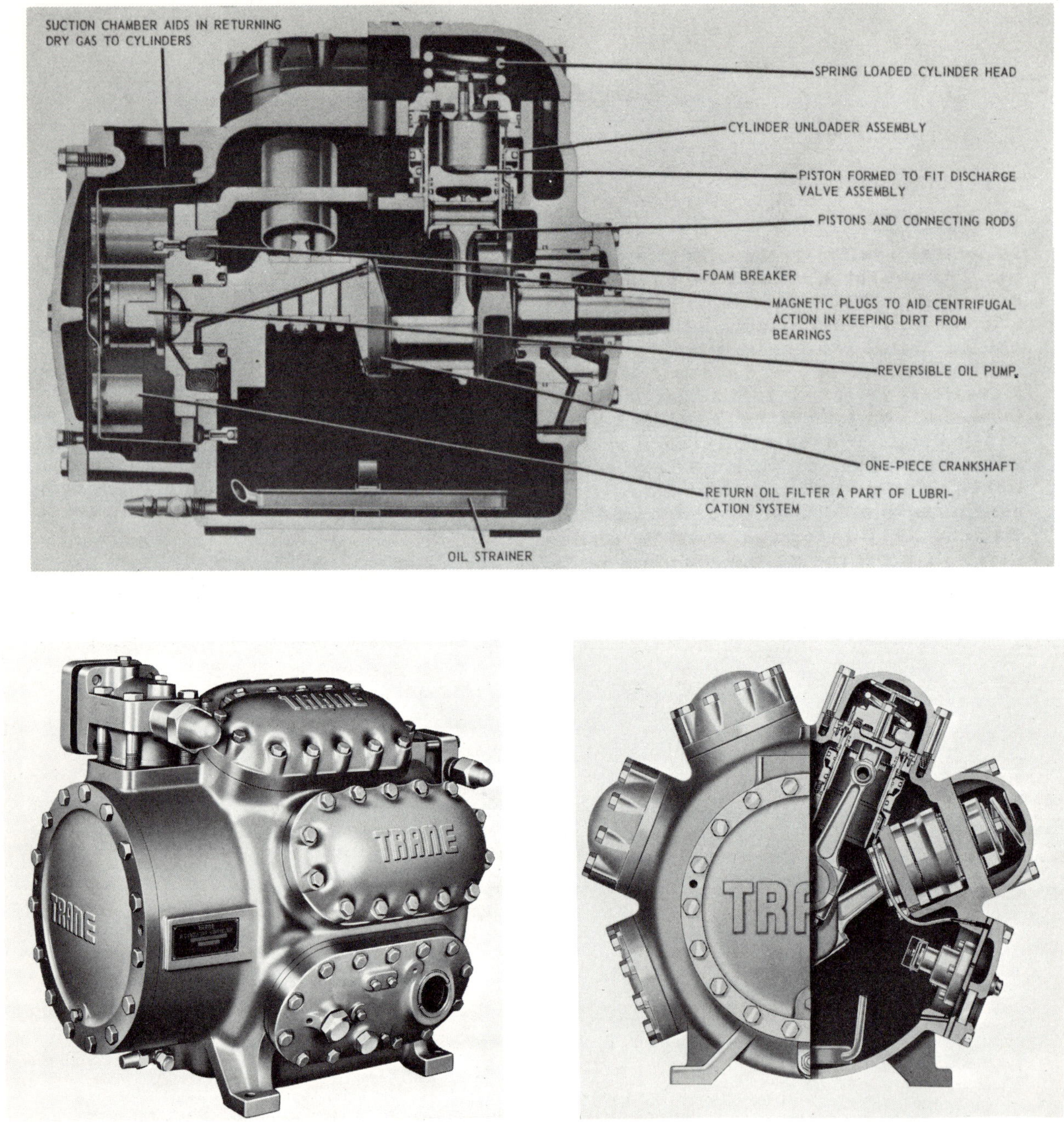

Courtesy of The Trane Company

Fig. 3-1. Reciprocating compressor (open type, showing components, explained above).

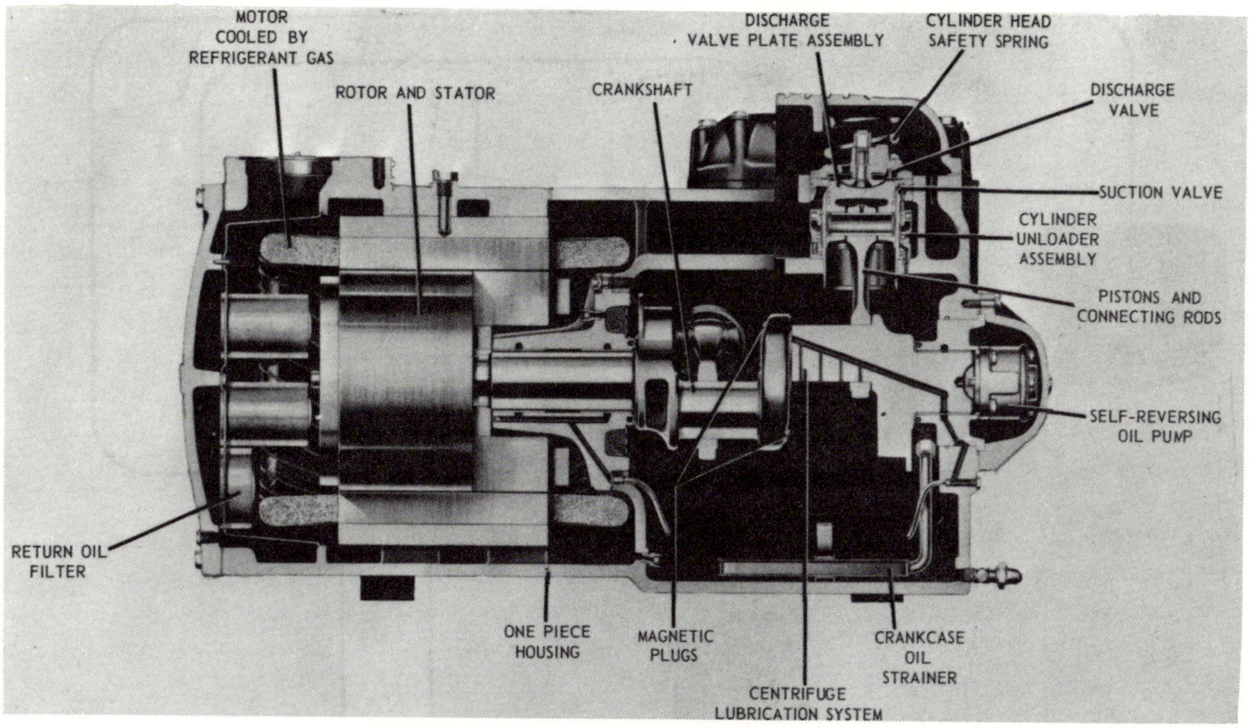

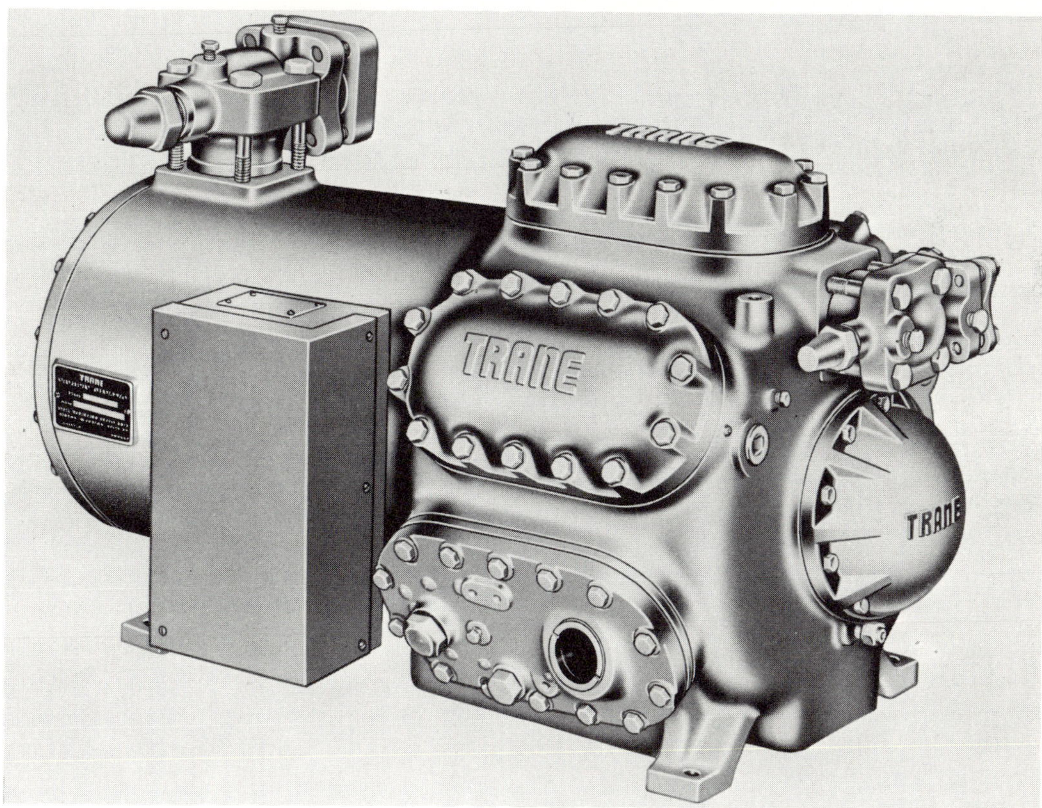

Courtesy of The Trane Company

Fig. 3-2. Reciprocating compressor (Hermetic type) showing components.

Courtesy of the Worthington Air Conditioning Company

Fig. 3-3. Compressor capacity control. A typical "unloading mechanism" which controls the capacity of a reciprocating compressor in accordance with the system demand. The method used in this case activates or deactivates cylinders by valving control of oil to or from cylinder unloaders.

is cooled by the refrigerant gas. The reciprocating type normally does not exceed 125 to 150 tons.

Centrifugal — The centrifugal compressor differs from the reciprocating type in that it uses the turbine principle. The impeller (or impellers where more than one stage is utilized) picks up the refrigerant gas and forces it into the condenser at pressure generated by the impeller. Figures 3-4 and 3-5 show cutaway views of this type of compressor. Figure 3-6 shows the refrigeration cycle. Figure 3-7 shows the interlocked controls to assure a safe start.

Condenser — The condenser lowers the high-temperature refrigerant gas (received from the compressor) to the point where the gas is condensed and delivered, as a high-pressure liquid, to the thermal expansion valve ahead of the chiller.

Chiller (*Evaporator or Cooler*) — The chiller receives the high-pressure liquid from the thermal expansion valve. By expanding in this low pressure area, the liquid boils, evaporates, and again becomes a gas.

Refrigeration Machine Cycle

Figure 3-6 shows how the compressor, condenser, and chiller (or cooler) are used in making up the refrigeration machine cycle.

Chilled-Water Cycle

Figure 3-8 shows the path of the chilled water as it leaves the chiller, how it is controlled, and how its temperature changes as it leaves the air handlers and is returned to the chiller.

Refrigerants

Properties — In developing refrigerants, many chemicals have been tried, early experiments being made with chloroform, ether, air, butane, and propane. The more successful refrigerants have been ammonia, methyl chloride, carbon dioxide, and sulfur dioxide. Ammonia, one of the oldest refrigerants, is still used extensively in ice plants and cold-storage warehouses. It has the disadvantage of being irritating, particularly to the eyes and mucous membranes. Of all refrigerants, carbon dioxide stands alone in being inert and noncorrosive. This is offset, however, by its high operating pressures, approximately 500 pounds on the low side and 1,000 pounds on the high side. By comparison, ammonia can be evaporated at 45 pounds and condensed at 200 pounds, and consequently, operates with much lighter equipment than that required by carbon dioxide. Refrigerant R-12, a comparatively recent member of the refrigerant family, functions at even lower pressures (approximately 30 pounds on the low side, and 135 pounds on the high side). The halogen group of refrigerants, Refrigerant R-11, Refrigerant R-12, and Refrigerant R-22 in particular, are noncorrosive, nonflammable, nontoxic, nonexplosive, colorless, and almost odorless. Each of these refrigerants has a different boiling temperature and different pressure differentials. Generally, Refrigerant R-12 and Refrigerant R-22 are used with reciprocating compressors while Refrigerant R-11 is used with centrifugal compressors which require a low-pressure refrigerant. Although the pressure of Refrigerant R-22 is different from that of Refrigerant R-12, handling is similar.

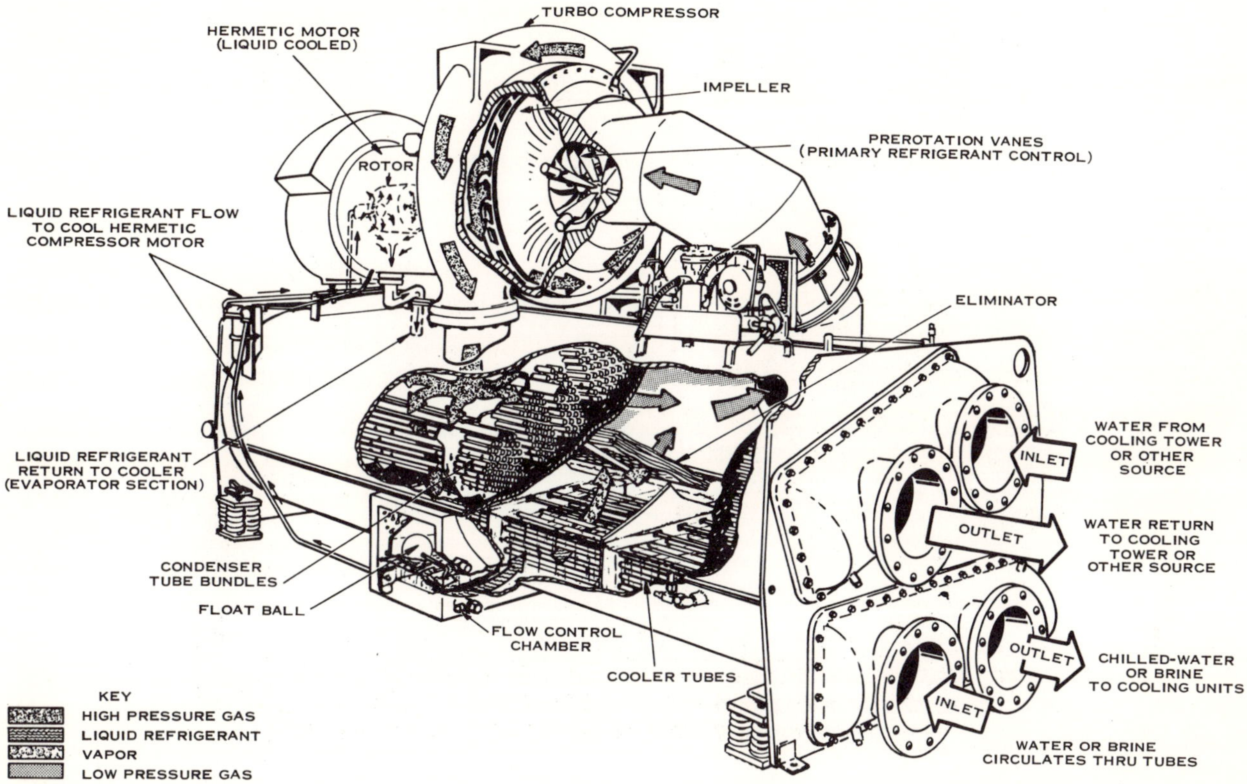

Courtesy of The York Corporation

Fig. 3-4. Centrifugal compressor installation showing cutaway view of components.

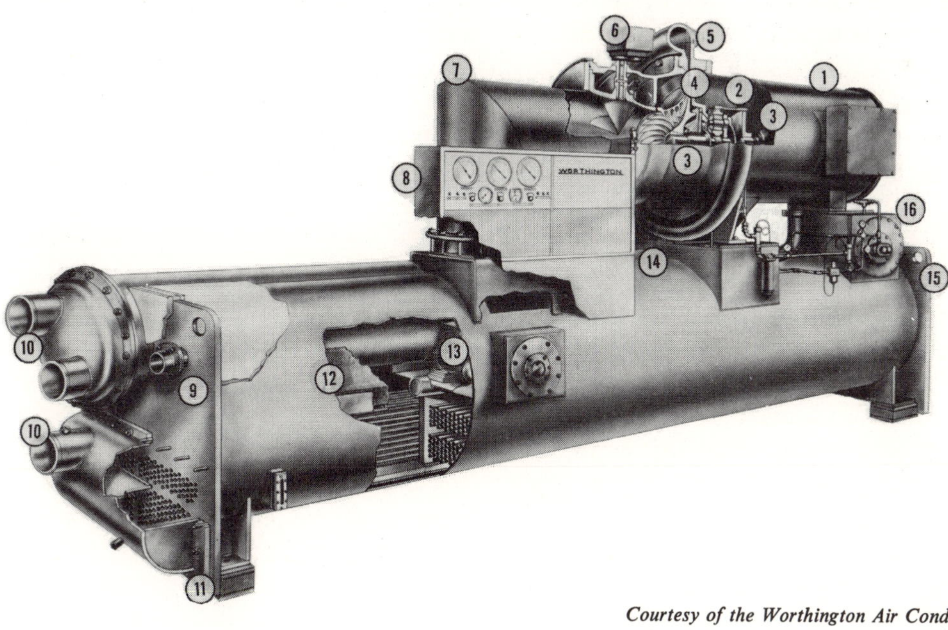

Courtesy of the Worthington Air Conditioning Company

Fig. 3-5. Centrifugal compressor and components: (1) Refrigerant-cooled motor; (2) Gears; (3) Bearings; (4) Impeller; (5) Volute; (6) Capacity-control assembly; (7) Suction elbow; (8) Control console; (9) Rupture disc; (10) Water connections; (11) Vibration isolators; (12) Eliminator; (13) Float valve; (14) Purge unit; (15) Lifting points; and (16) Lubrication system.

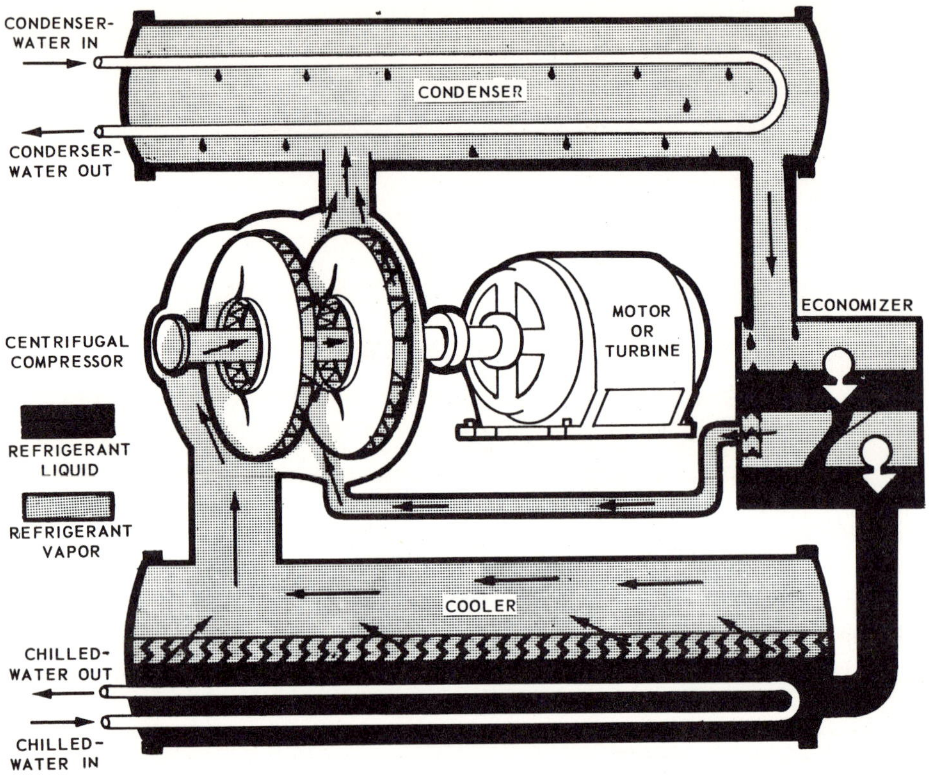

Courtesy of the Carrier Corporation

The operating cycle of the CENTRIFUGAL REFRIGERATING MACHINE is basically the same as in almost all refrigerating machines except for the method of compression. The refrigerant vapor coming from the cooler goes into an opening around the hub of the first wheel. The blades in this rapidly rotating wheel impart velocity to the vapor and centrifugal force throws it outward from the wheel into aerodynamically-designed passageways which reduce the speed of the vapor and convert the energy of velocity into pressure. The vapor is then directed into the hub of the second wheel where it is compressed still more before going to the condenser.

Between the condenser and the cooler, the liquid refrigerant passes through an economizer. By connecting this economizer to the second stage of the compressor, the pressure in the lower chamber is greatly reduced. This causes some of the liquid refrigerant to flash into vapor and cools the remainder by evaporation. Thus, the economizer acts as an interstage flash-cooler to increase the efficiency.

In the cooler, the liquid refrigerant changes its state to a vapor, absorbing heat from the brine or water, and, in doing so, supplies the refrigeration.

Fig. 3-6. Centrifugal refrigeration cycle.

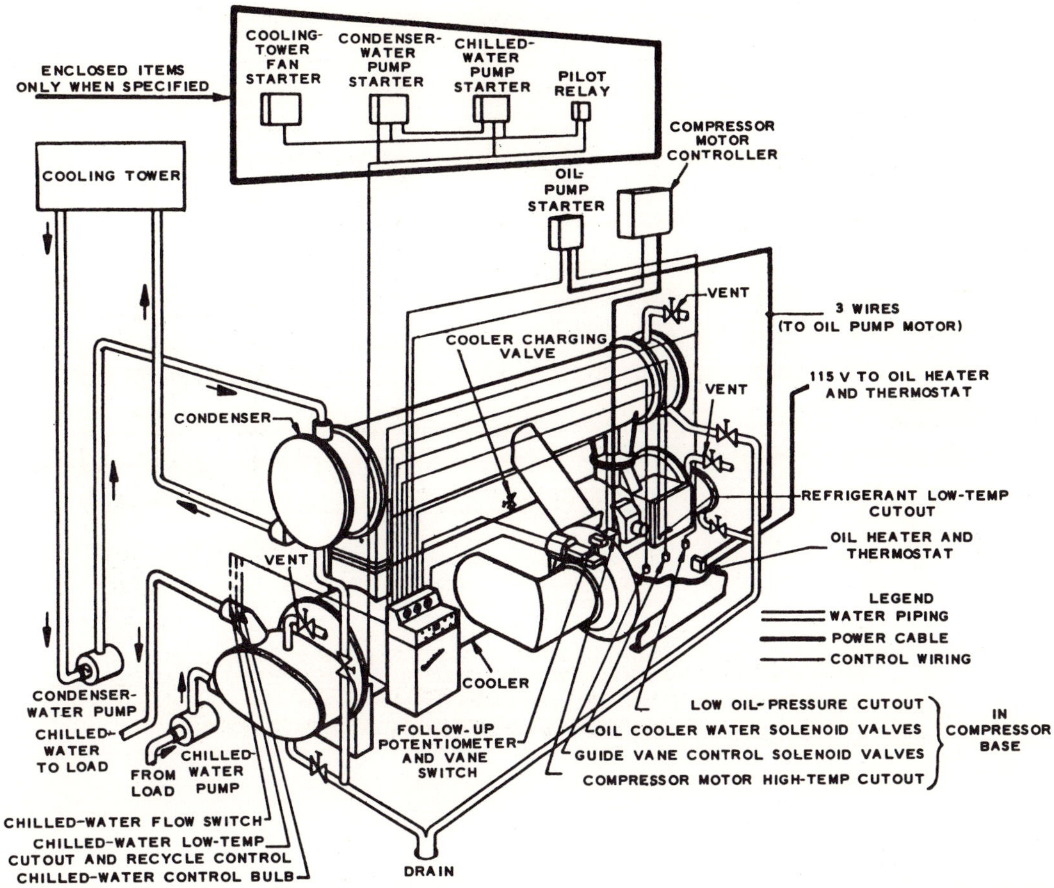

Courtesy of the Carrier Corporation

1. CHILLED-WATER PUMP—Pushing the "START" button on the compressor control panel, starts the chilled-water circulating pump. This automatically brings into action the relay activated by the return chilled-water thermostat.
2. CONDENSER WATER PUMP—If the return chilled-water temperature is at or above the cut-in or set point, the control relay will start the condenser-water pump.
3. TOWER FAN—The tower fan will, or will not operate, depending on whether the condensing-water temperature at the time is above or below control-point temperature.
4. COMPRESSOR MOTOR—If, at this point, all protective controls have been energized and have been satisfied, the compressor motor will start.

Fig. 3-7. Starting sequence of centrifugal compressor.

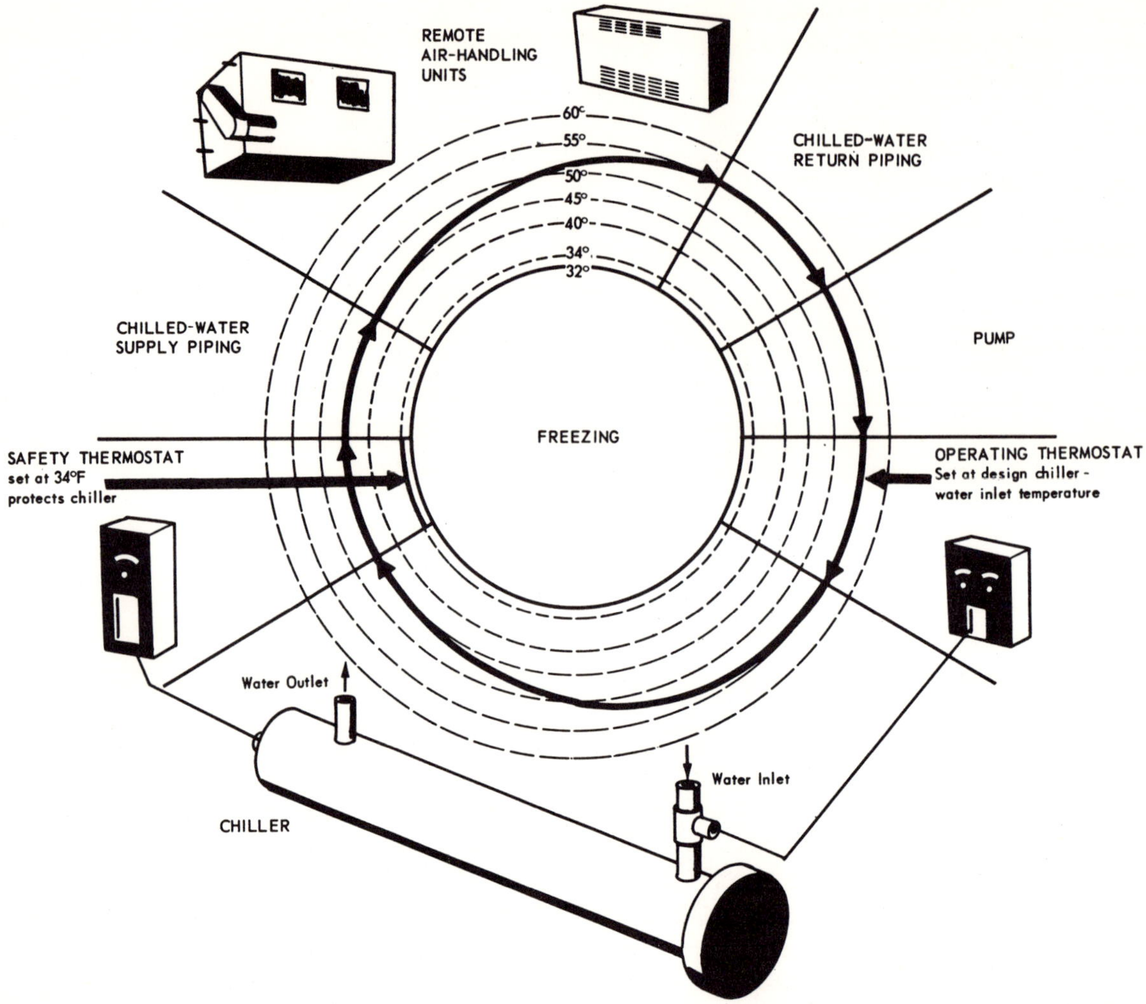

Courtesy of the Curtis Division of Transairco, Inc.

The SAFETY THERMOSTAT is mounted on the unit frame at the chiller-liquid outlet side, and the bulb is factory-mounted in an insertion well in that end of the chiller. For chilled-water applications this thermostat is normally factory-set at 34 deg. F and must not be changed. Notice how the temperature of the chilled water gradually rises from 45 deg. F as it leaves the chiller, to 55 deg. F. as it leaves the air-handling units. From here, the water is pumped back to the chiller where the temperature is reduced to 45 deg. F by the refrigeration machine, and is again discharged at this temperature. The rise from 45 deg. F to 55 deg. F represents the heat extracted from the air blown over the chilled-water coils in the air-handlers.

Fig. 3-8. Temperature graph of chilled-water circuit in typical chiller installation.

CAUTION: Whenever refrigerants are used, adequate ventilation should be provided because the accidental discharge of large quantities of gas, even though non-toxic, may produce a suffocating effect. Persons working in areas where there is a high concentration of certain refrigerants, such as methyl chloride and ammonia, must wear approved gas masks.

Summary of Refrigeration

Compressor

1. The compressor raises the pressure of the refrigerant gas from the evaporator pressure to the condensing pressure of the gas.
2. In so doing, it also raises the temperature of the gas.

Condenser

1. The condenser receives the high-pressure, high-temperature gas.
2. As the gas travels through the condenser it is condensed (changed to a liquid) by becoming cooled below its condensing temperature.
3. The condensing medium (air or water) must always be at a lower temperature than the refrigerant it is condensing.
4. In an air-cooled condenser, air is the condensing medium.
5. In a water-cooled condenser, water is the condensing medium.
6. In an evaporative condenser, both air and water are the condensing media. The condenser coils are wetted with water and then, by passing air over the wetted coils, some of the water evaporates. The refrigerating effect is accomplished by this evaporation.

Expansion Valve or Metering Device

1. This device provides the difference in pressure in the system.
2. On the condenser side of this device is high-pressure liquid, while on the evaporator side is low-temperature gas.

Evaporator (Chiller or Cooler)

1. The high-pressure liquid refrigerant is metered into the evaporator through the expansion valve or metering device.
2. Upon entering the evaporator, the pressure of the refrigerant drops below that at which it boils.
3. In the process of boiling, the refrigerant evaporates and becomes a gas. In other words, in the attempt to reach its low evaporative temperature, the drops of liquid must expand. In order to expand, they require heat.
4. In order to evaporate, the liquid must absorb heat and it gets this heat from the surrounding medium, either air or water.
5. It thus again becomes a gas and is ready to re-enter the compressor as a low-temperature, low-pressure gas.

Troubleshooting Guide

A typical troubleshooting guide for reciprocating machines has been provided in Table 3-1. This general guide should be used where the manufacturer of the machines did not furnish one. Most of the items are also applicable to centrifugal machines, but more complete guides for this type of machine should be available from the manufacturer.

Machine Room Record

A written record should be kept of machine performance. Usually, the manufacturer includes a log format in his maintenance manual. The log form records data important to the machine's operation and maintenance. Good interpretation of this data insures proper operation of the refrigeration plant and will help in avoiding costly breakdowns. Two different record forms are provided herein; one for reciprocating-type machines and one for centrifugal-type machines.

a. *Reciprocating-type Machine Form* — See Exhibit 3-A.
b. *Centrifugal-type Machine Form* — See Exhibit 3-B. NOTE: For large refrigeration plants using the Refrigeration Plant

Table 3-1. Trouble Diagnosis Chart

Complaint	Possible Causes	Corrective Steps
1. Compressor will not run	a) Main switch opens.	a) Close switch.
	b) Fuses blown.	b) Check electrical circuits and motor winding for shorts or grounds. Investigate for possible overloading. Replace fuse after fault is corrected.
	c) Thermal overloads tripped or open.	c) See corrective steps for open motor overload relays.
	d) Defective starter.	d) Repair or replace.
	e) System shut down by safety devices.	e) Determine type and cause of shut-down and correct it before resetting safety switch.
	f) Thermostat set too high.	f) Check evaporator temperature. Lower thermostat setting, if possible without freeze-up.
	g) Liquid line solenoid will not open.	g) Repair or replace.
	h) Motor electrical trouble.	h) Check motor for opens, short circuit, or burn-out.
	i) Loose wiring.	i) Check all wire junctions. Tighten all terminal screws.
2. Compressor noisy or vibrating	a) Improper isolation	a) Check isolator height.
	b) Gas pulsations.	b) Install muffler in discharge line.
	c) Coupling loose or misaligned.	c) Tighten bolts. Realign.
	d) Flooding of refrigerant into crankcase.	d) Check rating and setting of expansion valve.
	e) Improper piping support.	e) Relocate, add, or remove hangers.
	f) Improper clearances.	f) Running gear worn. Overhaul compressor and replace defective parts.
3. High discharge pressure	a) Condenser water insufficient or temperature too high.	a) Readjust water regulating valve. Investigate ways to increase water supply.
	b) Fouled condenser tubes (water-cooled condenser). Clogged spray nozzles (evaportive condenser). Dirty tube and fin surfaces (air-cooled condenser).	b) Clean.
	c) Non-condensibles in system.	c) Purge the non-condensibles.
	d) System overcharged with refrigerant.	d) Remove excess.
	e) Discharge shut-off valve partially closed.	e) Open valve.
	f) Condenser undersize.	f) Check condenser rating tables against the operation.
4. Low discharge pressure	a) Faulty condenser temperature regulation.	a) Check condenser control operation.
	b) Suction shut-off valve partially closed.	b) Open valve.
	c) Insufficient refrigerant in system.	c) Check for leaks. Repair and add charge.
	d) Low suction pressure.	d) See corrective steps for low suction pressure below.
	e) Compressor operating unloaded.	e) See corrective steps for failure of compressor to load up below.
	f) Condenser too large.	f) Check condenser rating table against the operation.
	g) Worn piston rings. Worn discharge valve service.	g) Overhaul compressor.
5. High suction pressure	a) Excessive load.	a) Reduce load or add equipment.
	b) Expansion valve over-feeding.	b) Check remote bulb. Regulate superheat. Check valve rating against the application.
	c) Compressor operating unloaded.	c) See corrective steps below for failure of compressor to load up.
	d) Compressor undersize. Evaporator too large.	d) Check ratings.

Courtesy of the Worthington Air Conditioning Company

Table 3-1 (Continued). Trouble Diagnosis Chart

Complaint	Possible Causes	Corrective Steps
6. Low suction pressure	a) Lack of refrigerant. b) Evaporator dirty or iced up. c) Clogged liquid line filter-drier. d) Clogged suction line or compressor suction gas strainers. e) Expansion valve malfunctioning. f) Condensing temperature too low. g) Compressor will not unload. h) Evaporator fan or chiller pump-off.	a) Check for leaks. Repair and add charge. b) Clean or defrost. c) Replace cartridge(s). d) Clean strainers. e) Check and reset for proper superheat. Repair or replace if necessary. f) Check means for regulating condensing temperature. g) See corrective steps for failure of compressor to unload. h) Check. Add interlock.
7. Compressor will not unload	a) Oil pressure above 65 psi. b) Clogged capacity control metering orifice. c) Defective capacity control. d) Unloader mechanism stuck. Unloader lifting pins worn. e) Auxiliary unloading control mechanism inoperative.	a) Clean oil pump relief valve. b) Clean orifice. c) Replace. d) Replace cylinder liner-unloader assembly. e) Repair or replace mechanism.
8. Compressor will not load up	a) Insufficient oil pressure. b) Damaged capacity control bellows. c) Defective capacity control. d) Defective unloader assembly. e) Auxiliary unloading control mechanism inoperative.	a) See corrective steps below for low oil pressure. b) Replace capacity control. c) Replace. d) Replace cylinder liner-unloader assembly. e) Repair or replace mechanism.
9. Compressor loading-unloading intervals too short	a) Oversize expansion valve causing excessive suction pressure. b) Erratic control oil pressure.	a) Check expansion valve tables. b) Check compressor oil level; foaming. Check oil pump relief valve.
10. Little or no oil pressure	a) Low oil pressure. b) Excessive liquid in crankcase. c) Loose fitting on oil suction strainer tube. d) Pump housing gasket installed improperly. e) Oil pressure gauge defective. f) Defective oil pump relief valve. g) Oil pump reversing gear stuck in wrong position. h) Low-oil pressure safety switch defective. i) Worn oil pump. j) Broken oil pump tang. k) Clogged suction oil strainer. l) Worn bearings. m) Worn unloader assembly. n) Broken control oil tube between capacity control and unloader.	a) See corrective steps below for loss of oil. Add oil. b) Energize crankcase heater. Reset expansion valve for higher superheat. Check liquid line solenoid valve operation. c) Check and tighten. d) Check. All holes in the housing face must be gasketed. e) Repair or replace. Keep valve closed except when taking readings. f) Repair or replace. g) Reverse direction of compressor rotation. h) Repair or replace. i) Replace. j) Replace pump assembly. k) Clean. l) Overhaul compressor. m) Replace cylinder liner-unloader assembly. n) Repair tubing.
11. Compressor loses oil	a) Clogged return oil strainer in suction end cover. b) Lack of refrigerant. c) Velocity in risers too low.	a) Replace. b) Check for leaks and repair. Add refrigerant. c) Check riser sizes.

Courtesy of the Worthington Air Conditioning Company

Table 3-1 (Concluded). Trouble Diagnosis Chart

Complaint	Possible Causes	Corrective Steps
	d) Oil trapped in line or evaporator.	d) Check pitch of lines and refrigerant velocities.
	e) Excessive compression ring blow-by	e) Overhaul compressor.
	f) Leaky unloader assembly.	f) Replace cylinder liner-unloader assembly.
12. Motor overload relays open	a) High condensing temperature.	a) See corrective steps for high discharge pressure.
	b) One line-fuse open, causing single-phase operation.	b) Find cause of fuse opening, repair, and replace fuse.
	c) Low voltage during high load conditions.	c) Check supply voltage and for excessive line drop.
	d) Loose power wiring.	d) Check all connections and tighten.
	e) Wrong size overload.	e) See motor manufacturer's data for proper size.
	f) Defective relay assembly.	f) Repair or replace.
	g) Defective or grounded wiring in motor or power circuits.	g) Repair or rewire.
	h) Power line fault causing single-phase running or unbalanced voltage.*	h) Check supply voltage. Notify power company. Do not restart until fault is corrected.
	i) High ambient temperature around the overload relay.	i) Provide cooling. Blow air over switch enclosure to reduce heat.
	j) Failure of second starter to pull up on part-winding systems.	j) Repair or replace starter or time delay mechanism.
	*On wye-delta power systems, if a primary leg of the transformer opens, serious voltage unbalance will occur in the secondary. When this failure occurs so that high current flows in the unprotected leg of the three phases, motor burn-out will occur. On across-the-line start motors, there is at least one overload in each phase, so that the seriousness of the condition is reduced. No remote overload protection can be provided, however, to protect fully under both single-phasing and unbalanced voltage conditions.	
13. Compressor thermal protector switch opens	a) Operating beyond design limits.	a) Check the table of limits. Add facilities so that conditions are within allowable values.
	b) Discharge valve partially shut.	b) Open valve.
	c) Blown valve plate gasket.	c) Replace gasket.
	d) Open relief valve	d) Replace.
	e) Broken discharge valve.	e) Replace valve service. Check compressor for additional damage.

Courtesy of the Worthington Air Conditioning Company

Operation Log (Exhibit 12-B, referred to in Chapter 12), duplicated items should be omitted from the Exhibit 3-B record form.

Seasonal Start-up

Checks Prior to Start-up — The following procedures, although general in nature, should, if properly carried out, result in a trouble-free start.

System — Check the component parts of the refrigeration cycle and the system, both visually and manually, to determine that all valves and mechanical operators are correctly positioned for the system's mechanical operation.

Compressor — Check oil level and refrigerant level.

Condensing-Water System — Check the cooling tower and fill the condensing system. (It is, of course, assumed that the tower had been properly cleaned after the preceding cooling season and cleaned once again, prior to filling the condensing-water system.) Check to see if bleed line is tapped into condensing-water line between pump and the tower so that water is bled to waste only when the tower pump is operating. Add necessary water-treatment chemicals.

Chilled-Water System — Check expansion tank, if provided, for chilled-water level.

Operation — When it has been made certain that all of the foregoing checks have been satisfactorily completed, the plant is then ready for start-up. In the case of installations which include

REFRIGERATION RECORD
(Reciprocating Type Machine)

BUILDING	MONTH AND YEAR

MACHINE NUMBER AND DATE	TIME CHECKED	OUTSIDE CONDITIONS				COMPRESSOR									CONDENSER		CHILLER	
		TEMPERATURES		SUN-SHINE (Check one) (5)		HIGH SIDE HEAD		LOW SIDE SUCTION		OIL GEAR BOX					WATER TEMPERATURE		WATER TEMPERATURE	
		D.B.	W.B.			PRES-SURE	TEMPER-ATURE	PRES-SURE	TEMPER-ATURE	PRES-SURE	TEMPER-ATURE				IN	OUT	IN	OUT
(1)	(2)	(3)	(4)	YES	NO	(6)	(7)	(8)	(9)	(10)	(11)	(12)	(13)	(14)	(15)	(16)	(17)	(18)
	____ A.M.																	
____	____ A.M.																	
(No.)	____ P.M.																	
____	____ P.M.																	
(Date)	____ P.M.																	
	____ P.M.																	
	____ A.M.																	
____	____ A.M.																	
(No.)	____ P.M.																	
____	____ P.M.																	
(Date)	____ P.M.																	
	____ P.M.																	
	____ A.M.																	
____	____ A.M.																	
(No.)	____ P.M.																	
____	____ P.M.																	
(Date)	____ P.M.																	
	____ P.M.																	
	____ A.M.																	
____	____ A.M.																	
(No.)	____ P.M.																	
____	____ P.M.																	
(Date)	____ P.M.																	
	____ P.M.																	

INSTRUCTIONS

GENERAL—Where 1 machine is installed, use the form to cover 4 days operation; where 2 machines are installed, use the form to cover 2 days operation; and, where 3 or 4 machines, use to cover each day of operation.

COLUMN (1)—Show the number of the machine in operation and the date. If machine is not in operation, so indicate.

COLUMN (2)—Readings are to be taken at the end and middle of each tour.

COLUMN (11)—Applies to open units only where they have gear box.

COMMENTS—Use reverse for comments or notes deemed necessary or of interest.

Exhibit 3-A. Refrigeration record form for reciprocating type machine.

BUILDING			REFRIGERATION RECORD (Centrifugal Type Machine)																				MONTH AND YEAR			
							COOLER					CONDENSER				COMPRESSOR							THERMAL PURGE			
DATE	MA-CHINE NUMBER	TIME CHECK-ED	OUTSIDE CONDITIONS				GPM ______					GPM ______				Position Cap. Indicator	BEARING TEMP.		OIL				Purge Pressure for Pump Operation		Time Be-tween Pump Starts	Refrig. Level
			TEMPERA-TURES		SUN-SHINE (Check one)		Vacuum	Liquor Temp.	Liquid Level	WATER TEMP.		Vacuum or Press.	Cond. Temp	WATER TEMP.			Main	Motor End	Level	Temp.	Pressure	Motor Amps	Starts	Stops		
																							COMPRESSION PURGE			
			D.B.	W.B.	YES	NO				In	Out			In	Out								Suct. Press.	Disch. Press.	Oil Level	Refrig. Level
1	2	3	4	5	6	7	8	9	10	11	12	13	14	15	16	17	18	19	20	21	22	23	24	25	26	27

COMMENTS

Exhibit 3-B. Refrigeration record form for centrifugal type machine.

water chillers and cooling towers, the operator should become familiar with Fig. 3-7, which explains the sequence of operations that takes place upon pushing the START button. These built-in safety controls could also explain the reason for some non-starts.

> CAUTION: In starting a refrigeration plant after a prolonged shut-down where the indoor and outdoor temperatures and humidity are relatively the same — *do not attempt to decrease* the inside temperature *too rapidly*. The refrigerating equipment should be operated for several hours at reduced capacity, slowly lowering the chilled-water temperature as the temperature and dewpoint of the air in the building are gradually reduced.

Start-Up Procedures — The following procedures should be carried out at the beginning of the cooling season or directly after a prolonged shut-down in daily operation. All procedures shall conform with manufacturer's recommendations. (See correct practice, in Condensing-Water System, above.)

1. Turn oil pump on, if equipped with manual control. Operate for 10 minutes before starting the unit. If not equipped with a thermostat, feel the oil sump which should be warm to the touch.
2. Centrifugal Machines — Start purge pump and operate for 10 minutes before starting unit.
3. Check refrigerant level. Check oil level; should be anywhere in the sight glass, but normally, about half-way.
4. Set all valves in the chilled-water and condenser-water lines to correct position for operation. Make certain all summer and winter change-over valves and switches have been placed in proper position and are not subsequently changed by operating personnel.
5. Check all refrigerant line valves for proper scttings.
6. If there is more than one compressor: (a) Energize correct chilled-water and condenser-water pumps; (b) De-energize pumps serving idle compressors.
7. Start unit if oil temperature is up to the required point.
8. For the following 30 minutes, continually check oil and refrigerant levels. Oil spots, air bleed (from the purge valve in centrifugal machines), and bubbles in sight glass are indications of possible leaks. Hunting on start-up of centrifugal machines indicates a possible leak.
9. If necessary, manually control capacity of machines to avoid cutting-out on overload until building temperature is within range for the machine to handle the load automatically.

> CAUTION: In doing this, the operator must be certain that the machine is not cutting-out due to a malfunction within the machine or in the system.

Operational Checks

During the cooling season, the following checks should be made and where applicable, shall conform with manufacturer's recommendations.

Refrigeration Plant

1. Observe oil level in "bull's eye."
2. Check flow of refrigerant through sight glass. Should be smooth and without bubbles. (Where sight glass has sealed caps, the caps should be removed to make the check. When replacing the caps, make them hand tight. Do not use wrench.)
3. Observe any unusual noise of machine operation.
4. Observe any visual changes in equipment appearance such as oil spots on connections, or on the floor under equipment.
5. With leak detector, check for leaks of refrigerant and oil around shaft seal, sight glasses, valve bonnets, flanges, and flare connections, as well as at the relief valve on condenser assembly.
6. Check operating pressures and temperatures. Know the normal operating pressure

and temperature for each machine so that any variance from normal may be noted before malfunction is experienced.

7. Check operation of refrigeration controls and reduction devices; also the superheat setting of metering devices. (Inlet and outlet of evaporator.)
8. Check high-pressure cut-out setting which should be at least 25 pounds below the relief-valve setting.
9. When shutting down unit at any time, observe pressure cut-out point on low-pressure control.
10. Where applicable, lubricate motor bearings and all moving parts, as recommended by the manufacturer.
11. Check for necessary purging of air from system.
12. Check and adjust tension and alignment of all belts.
13. Check pump packing, and adjust where necessary.
14. Clean all strainers periodically.
15. Check all safety controls.

Condensing-Water System

1. Check level of water in cooling-tower sump. Adjust float valve if necessary.
2. Check to see if bleed is working properly. (Do not confuse with tower overflow line.) Bleed line is usually located in the condensing-water line between the pump and the tower.
3. Check for proper bleed, following procedures outlined in Chapter 5.
4. With unit shut down, check condenser-line strainer ahead of condenser-water pump. (Where unusual conditions exist, clean as often as required.)

Seasonal Shut-Down

Check all bolts on each piece of equipment, for tightness. Here, and in every procedure listed below, the manufacturer's recommendations should be followed.

Compressor

1. Shut off liquid line valve and allow machine to pump down. Valves that have been shut off should be "Red Tagged."
2. Check entire refrigeration system for refrigeration leaks.
3. For open-type machines, where there is qualified personnel, open the suction and discharge valves once a month during the off-season, and run unit a few revolutions to lubricate shaft seal. Re-close valves.

Electric Motors

1. Blow out motors with air under pressure, when available.
2. Employing qualified personnel, clean and check motor starting switches.
3. Have all electric leads checked for tightness of connections in all junction boxes and contactors, again using only qualified men.
4. Lubricate all motor bearings as recommended by manufacturers.

Condenser — If high or increasing condenser-water temperatures or pressures during the cooling season indicate the possible clogging of tubes, take action as follows:

1. *Condensers with Removable Water-Head* — Remove heads and examine for scale. If scale is present, clean with a rotary nylon brush. Never use a wire brush.
2. *Condenser with Non-removable Water-Head* — The tubes should be acid-cleaned by qualified personnel or under contract with a reputable company. There are two methods of doing this:
   (a) The condenser alone may be cleaned by an acid pump connected to the inlet and outlet of the condenser, circulating the cleaning solution from a container through the condenser tubes.
   (b) The tower basin can be utilized and the chemical solution circulated throughout the condensing-water system.

> CAUTION: Some systems are so interlocked electrically, that the condenser-water pump cannot be operated without energizing the compressor. In such a case, be sure that a hand-off-automatic control has been provided to allow for condenser-water pump operation when necessary for this procedure.

After cleaning and inspection, the condenser(s) should be filled with treated water to prevent corrosion. The water level should be kept low enough to prevent freezing.

Air-Cooled Condensers — Vacuum clean or wash condenser coils or fins of air-cooled condensers.

Chiller (Evaporator or Cooler) — Normally, chemical treatment for chillers is a one-time operation to bring the concentration of inhibitor up to a desired level. However, there may be a small loss of water each year through pump glands, possible overflow of the expansion tank, and drainage of parts of the system likely to freeze. If there is evidence of clogging, the procedure for cleaning condensers applies.

Pumps — At the end of the cooling season, check and observe operation of all pumps to determine what parts may be needed to put each pump in proper operating condition during shut-down in the off-season. Replace packing when adjustment has to be made too frequently, or when packing gland is screwed-in to the limit. Always check condition of shaft when packing is removed; if shaft sleeve is worn, it should be replaced before new packing is installed.

Miscellaneous — (a) Look for any signs of rust which will need touch-up painting, (b) List any needed repairs or replacement of worn-out parts.

Air-cooled Plant

With the growing need for conservation of water, there is an increasing effort on the part of manufacturers to produce air-cooled equipment that will handle larger cooling loads without the use of evaporative condensers or water-cooling towers. At present, packaged air-cooled refrigeration is available through the use of air-cooled condensers capable of providing as much as 60 tons of refrigeration, and this tonnage is increasing year-by-year.

Figure 3-9 shows one of these units. It is a completely packaged water-chiller designed for outdoor installation. It requires only a hook-up to power source and running the necessary chilled-water piping to the coils, in central air-distribution systems or to individual room fan-coil units. The use of air-cooled equipment such as this not only eliminates the need for water treatment and reduces water costs, it also permits operation in freezing weather without the modification and extra attention required by evaporative condensers and cooling-towers. This sounds like the complete answer. However, the problem of deciding which is the best method for a particular building is not that easy. It requires an economic study based on initial cost, investment cost, depreciation, fuel and utility costs, repair costs, and maintenance. This can become quite involved and requires the services of a qualified engineer to determine the most advantageous equipment and system for a particular building.

Prime Movers Other Than Electric

Absorption Chiller — Heat for operation is supplied from low-pressure steam or hot water. Water is considered to be the refrigerant in an absorption chiller.

Cycle of Absorption-chiller Operation

1. *Two Vessels* — It will be noted in Fig. 3-10 that the machine primarily consists of two vessels stacked vertically. The upper vessel contains the absorber and evaporator; the lower, smaller vessel contains the condenser, generator, and heat exchanger.
2. *Refrigeration Effect* — The refrigeration or cooling effect takes place in the evaporator.
3. *Cooling Cycle* — The refrigerant is sprayed by the refrigerant pump over a tube bundle in the evaporator section through which passes the water to be chilled. The refrigerant evaporates on the tube surfaces. The necessary heat required to evaporate the refrigerant is taken from the water in the

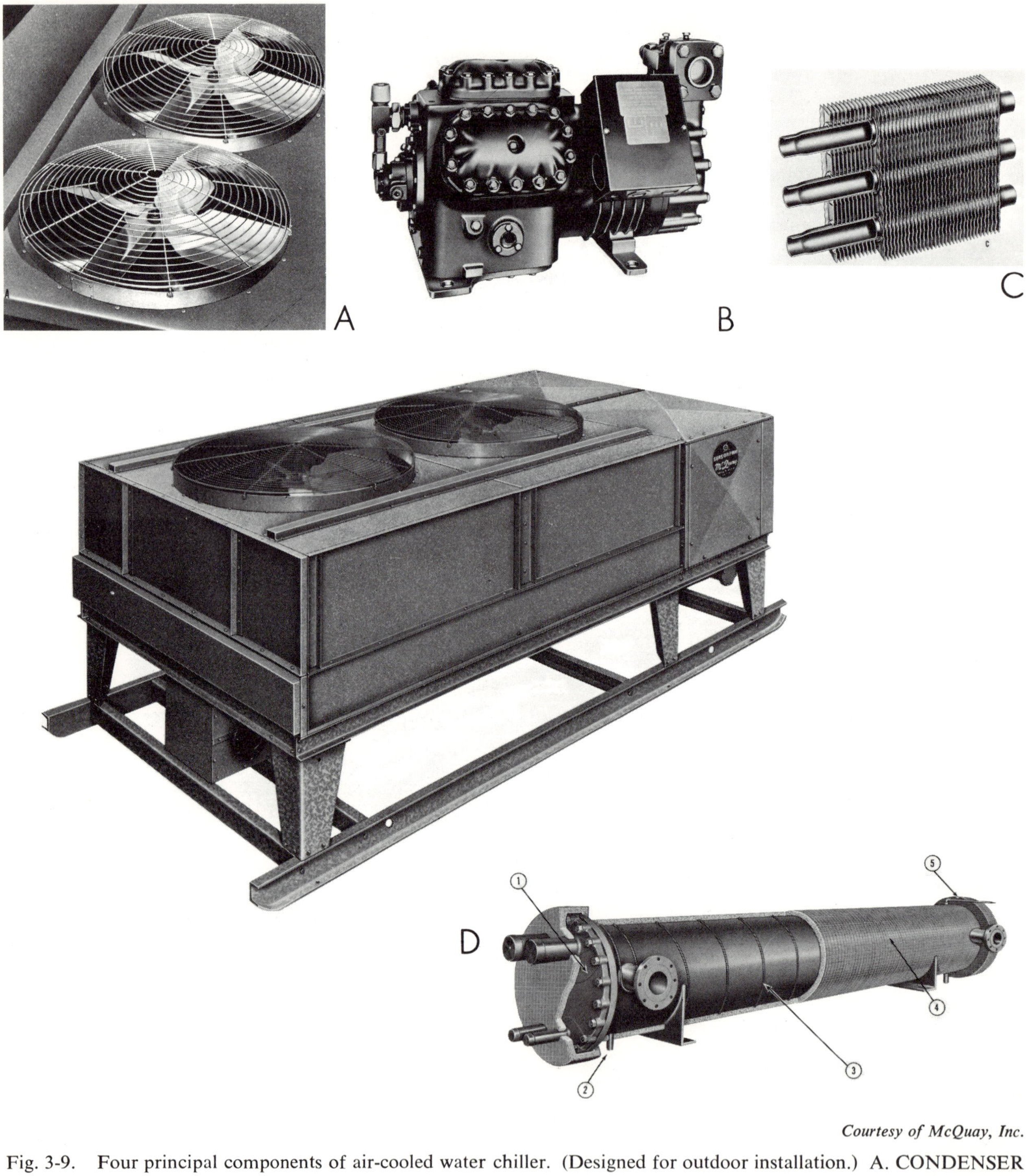

Courtesy of McQuay, Inc.

Fig. 3-9. Four principal components of air-cooled water chiller. (Designed for outdoor installation.) A. CONDENSER FANS—Heavy-duty fans are run at low tip speeds for quiet operation; B. ACCESSIBLE HERMETIC COMPRESSORS; C. HIGH-CAPACITY CONDENSER; D. (1) DIRECT-EXPANSION COOLER—Baffled shell and through-tube type with removable heads (connections reversed for clarity); (2) DRAIN CONNECTION—Provided at each end to facilitate draining; (3) HEATING CABLE—Automatically protects cooler down to $-20°F$. Complete factory-wired with control thermostat; (4) INSULATION—Heavy layer of foil-faced glass-fiber insulation with vapor seal. Prevents condensation on shell; (5) VENT CONNECTION—Piped and extended through cooler enclosure.

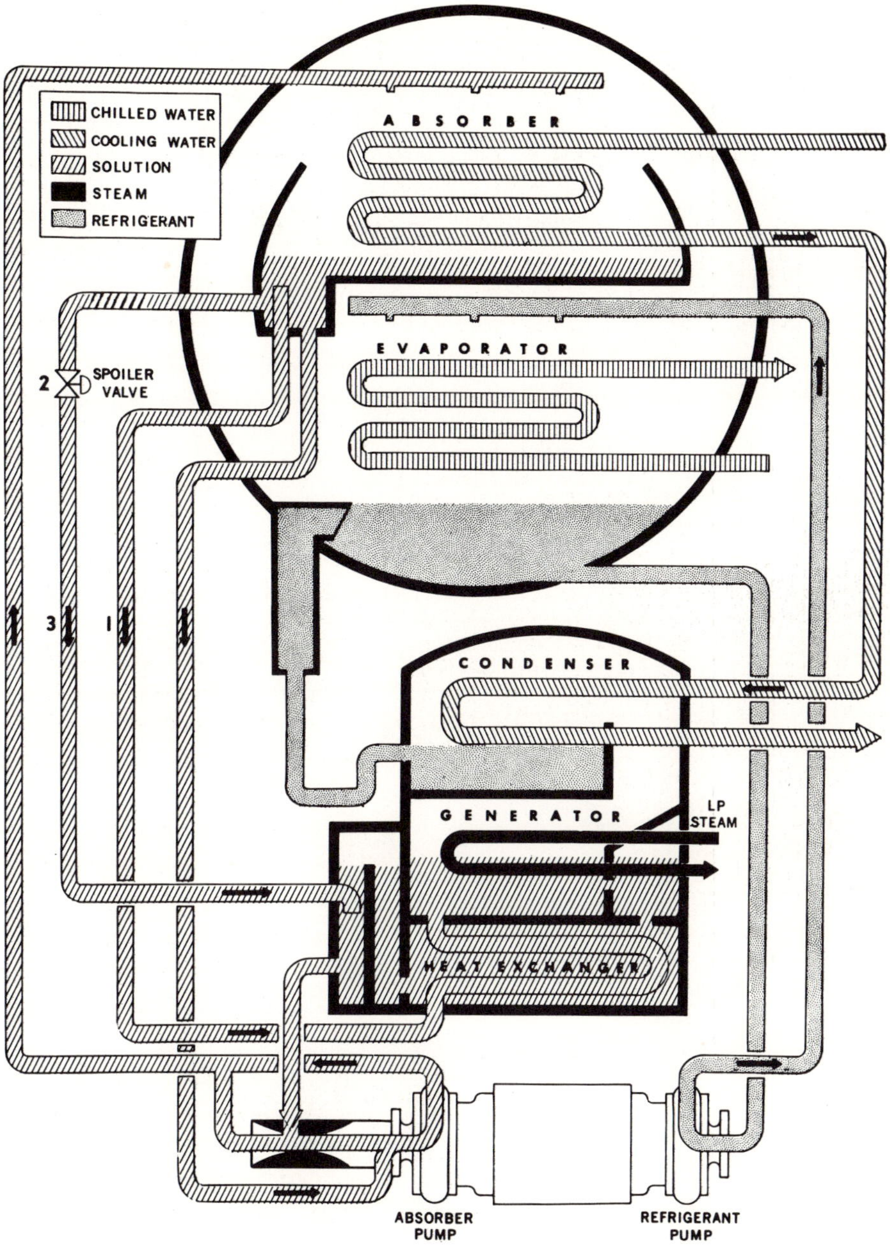

Courtesy of the Worthington Air Conditioning Company

Fig. 3-10. Diagram of absorption machine.

tubes, thereby chilling this water. The evaporator, in this case, is maintained at .25 inch of mercury. This pressure is low enough so that the boiling temperature of the refrigerant is sufficiently low to produce the desired leaving chilled-water temperature.

4. *Complete Cycle* — The complete cycle is extremely complex and will not be covered here.

5. *Vibration and Noise* — The vibration is practically nil and the noise level is very much less than that of the centrifugal.
6. *Electric Power* — The only electricity required is that for the two pumps, which is nominal.
7. *Capacity* — The maximum capacity for a single machine is 1,000 tons of refrigeration compared to 6,000 tons for a motor-driven centrifugal machine. On the average, the weight of an absorption machine exceeds that of a comparable capacity, electrically driven centrifugal, by from 25 to 50 per cent. The difference in weight increases directly with the capacity.
8. *Cooling Tower Capacity* — The cooling tower capacity required for an absorption machine is approximately twice that required by the centrifugal machine for the same tonnage capacity.
9. *Economics* — The initial equipment cost of an absorption machine could be from 10 to 15 per cent more than the motor-driven centrifugal. Depending on the electrical service available at an individual building, the difference in installed cost could be fairly comparable. To complete the comparative cost picture, the cost of electricity and all competitive fuels have to be taken into consideration, for the operating costs may offset the higher initial cost.

Gas, Diesel, or Turbine Drive — These engines, as primary movers, have their place in areas where electric rates are high or where electric service is not available. They also have an advantage where exhaust gases can be utilized as a source of heat.

Total Energy

In large buildings where the exhaust gases from primary movers can be utilized to the fullest extent, total energy has its place. In this case, electricity is generated through the use of gas or diesel-driven engines or through a gas turbine. The exhaust gases from the engines or turbines can be used to supply enough heat to both heat a building and run an absorption-type chiller. In those cases where the exhaust gases are not sufficient, a relatively small supplemental boiler can be used. The gas turbine provides the largest amount of exhaust gases and at high temperature. However, to justify its higher initial cost, the exhaust gases must be utilized to the fullest extent.

Economics — From the foregoing, it can be seen that the choice of a refrigeration plant for the larger buildings is not a simple matter. To decide which system is the most advantageous for a particular installation, the services of a qualified engineer should be utilized. He will make an economic analysis based on owning and operating costs, which is the only true way of determining the type plant and system which will result in the lowest annual owning and operating costs.

TERMS AND DEFINITIONS COMMON TO REFRIGERATION[1]

BOILING POINT OR TEMPERATURE — The temperature at which a fluid will change from a liquid to a gas. The boiling point will depend upon the pressure exerted on the surface of the liquid.

Btu OR BRITISH THERMAL UNIT — A unit of heat. The heat required to raise the temperature of one pound of water, at its maximum density, one degree Fahrenheit. Also, the heat to be removed in cooling one pound of water one degree Fahrenheit.

CAPILLARY TUBE — A small-diameter tube used between the bulb and power element of thermostatic expansion valves and remote bulb temperature controllers.

CENTIGRADE — A thermometer graduated or divided into 100 degrees. The 0° corresponds to 32° on the Fahrenheit scale and the 100° corresponds to the 212° on the Fahrenheit scale.

CHANGE OF STATE — Change of a liquid to a solid state upon freezing or to a gas upon evaporization, or vice versa.

[1] *Source:* "Refrigeration Manual," The Trane Co., (selected items from the Glossary).

CHARGE, OPERATING — The total amount of refrigerant required for proper operation.

COMPRESSION RATIO — The ratio obtained by dividing the discharge pressure in psi by the suction pressure in psi.

CUT-IN POINT — The pressure or temperature at which the equipment is set to start-up.

CUT-OUT POINT — The pressure or temperature at which the equipment is set to stop.

DEGREE DAY — Accepted standard is taken as being the difference between 65° and the mean dry-bulb temperature for a particular day. Example: For a mean dry-bulb temperature of 70 degrees, there would be 5 degree days.

DEHYDRATOR — A device containing a drying agent (desiccant), for removing moisture from the refrigerant.

DENSITY — The weight per unit-volume of a substance.

DESIGN TEMPERATURE, OUTDOOR — The outdoor temperature against which a system must be able to maintain the desired indoor conditions.

EFFICIENCY, OVER-ALL — Ratio obtained by dividing the theoretical power required, by the actual power required.

EVAPORATION — The conversion of a substance into a vapor.

EVAPORATOR, DIRECT EXPANSION — An evaporator where the medium to be cooled is in direct contact with the evaporator.

EXPANSION VALVE — The valve that meters the flow of liquid refrigerant to an evaporator.

FAHRENHEIT — A temperature scale where +32° is the freezing point and +212° is the boiling point of water, under standard atmospheric conditions.

FLASH GAS — Gas that is generated when the pressure on a liquid, held at the boiling point, is reduced.

GAGE, COMPOUND — A device that measures gas pressures both above and below atmospheric pressure.

HUMIDITY, RELATIVE — The amount of moisture in the air stated in terms of percentage of total saturation at the existing Dry-bulb temperature.

HUNTING — Fluctuations in the operation of a machine as the controls try to establish equilibrium; also the movement of liquid in a sight glass or calibrated glass pressure-gage.

LOW LIMIT CONTROL — A protective device to prevent operation below a predetermined set-point.

LOW PRESSURE CONTROL — A pressure-operated switch located in the suction side of a refrigeration or air-conditioning system that acts to stop the compressor at a fixed cut-out setting.

OIL, ENTRAINED — Drops of oil carried by high-velocity refrigerant gas.

OIL-PRESSURE FAILURE CONTROL — A device that shuts off a compressor whenever the oil pressure falls below a predetermined set-point.

OIL SEPARATOR — A device used to separate any oil entrained in the discharge gas from a compressor and return it to the compressor crankcase.

PRESSURE:

Atmospheric — Pressure exerted by the earth's atmosphere. At sea-level, under normal conditions, the atmospheric pressure is 14.7 *psi.*

Discharge — The pressure against which the compressor delivers the refrigerant gas.

Pressure Drop — The loss of pressure due to friction or lift.

Saturation — The pressure at which gas, at a specific temperature, is saturated.

Suction — The pressure forcing the gas to enter the suction inlet of the compressor.

PUMP DOWN — The reduction of pressure in a system.

PURGE — Discharge to atmosphere of impurities and noncombustible gases.

RECEIVER — That part of the machine that stores or retains the refrigerant liquefied by the condenser.

REFRIGERATION CAPACITY — Rate at which a system can remove heat; usually stated in tons, or Btu per hour.

REMOTE BULB — Controls the action of the expansion valve by assuming the temperature of the suction gas at the point where the bulb is fastened to the suction line. A change in the suction gas superheat at this point will operate the valve in a compensating direction to restore the superheat to a predetermined valve-setting.

SHORT CYCLE — Too-frequent starting and stopping. Short on-cycle and short off-cycle.

SIGHT GLASS, LIQUID — A glass "bull's-eye" installed in the liquid line permitting visual inspection of the liquid refrigerant; primarily for the purpose of detecting bubbles in the liquid which indicates a shortage of refrigerant in the system.

SIGHT GLASS, OIL — A glass "bull's-eye" in the compressor crankcase permitting visual inspection of the compressor oil-level.

SOAP-BUBBLE TEST — A method of detecting leaks by coating with a soap solution; any major leak will cause bubbles to form.

SOLENOID VALVE — A magnetically operated valve generally used to control the flow of liquid to an evaporator, but which may be used where-ever off-on control is permissible. A solenoid electrical winding controls the action of the valve.

SUBCOOLING — The cooling of liquid below its condensing temperature.

SUCTION HOLD-BACK VALVE — A regulating valve in the suction line designed to prevent the suction pressure at the compressor from rising above the predetermined setting of the valve. Used to prevent overloading the compressor.

SUCTION LINE — The pipe to conduct refrigerant vapor from the evaporator to the compressor.

SUCTION PRESSURE REGULATOR — An automatic valve or control device designed to maintain the pressure, and therefore the temperature in an evaporator, above a predetermined minimum.

SUPER HEAT — The temperature increase above saturation temperature or above boiling point.

TEMPERATURE:

Ambient — The temperature of the air surrounding the object under consideration.

Condensing — The temperature of the fluid in the condenser at the time of condensation.

Discharge — The temperature of the gas as it leaves the compressor.

Dry-Bulb — The temperature of air measured with an ordinary thermometer. It indicates only sensible heat changes.

Evaporating — The temperature at which the fluid boils under the existing pressure.

Saturation — The boiling point of a refrigerant at a given pressure. It is considered as the evaporator temperature in refrigeration.

Suction — The temperature of the gas as it enters the compressor.

Wet-Bulb — The temperature of the air measured with a thermometer having the bulb covered by a moistened wick.

THERMOMETER WELL — A small pocket, or recess in a pipe or tube, designed to provide good thermal contact with a test thermometer.

THROTTLING VALVE — A small valve used primarily in gage lines to shut off the line when readings are not to be made; and throttling the line to prevent fluctuations when readings are being made.

TON OF REFRIGERATION — A unit of refrigeration capacity corresponding to the removal of 12,000 Btu per hour. It is so-named because it is equivalent in cooling effect to melting one ton of ice in 24 hours.

VACUUM — A reduction in pressure below atmospheric pressure; usually stated in inches of mercury.

VAPOR — A fluid in the gaseous state following evaporation.

VAPORIZATION — Evaporation or boiling.

VENT — A port or opening through which pressure is relieved.

CHAPTER 4

Water-Cooling Towers

General

Where cooling towers are used, the water from the tower is circulated through the condenser, returned to the tower, then, after passing through the tower, it is recirculated through the condenser. (See Fig. 4-1.) The cooling tower conserves water and lowers the temperature of the water being circulated. In localities where the cost of water is not prohibitive, condensing-water systems are sometimes designed as a "once-through system," in which the cooling tower is eliminated and the condensing water is wasted directly to a sewer connection. Thus, cooling towers avoid waste. This can be appreciated when it is realized that the average cooling tower circulates approximately 3 GPM (Gallons per minute) per ton of refrigeration. A cooling tower system may be classed as an "open system," since it is open to atmosphere at the tower.

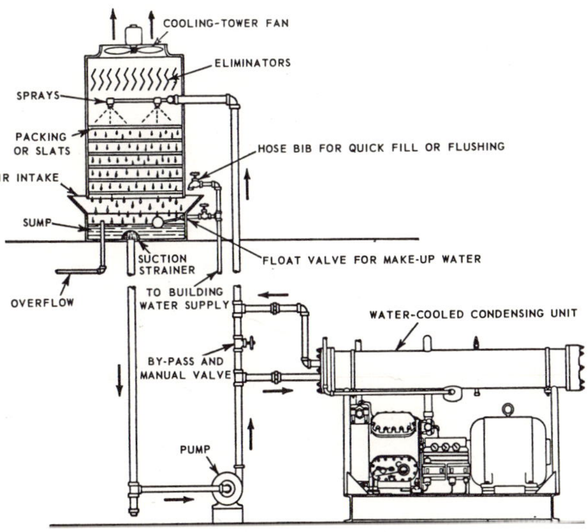

Courtesy of The Trane Company

Fig. 4-1. Relationship of cooling-tower, condenser, and compressor.

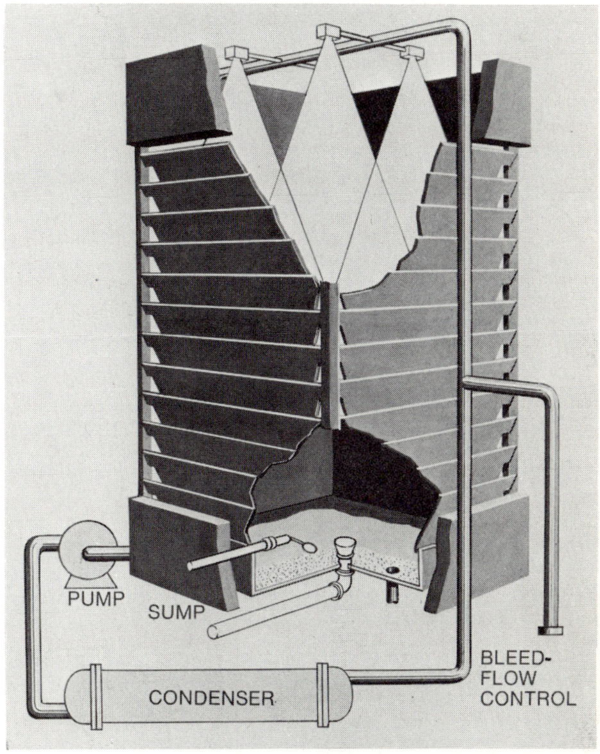

Courtesy of the Calgon Corporation

Fig. 4-2. Typical, small cooling tower.

Typical Cooling Tower — Shown in Fig. 4-2 is a typical cooling tower for a small installation, and in Fig. 4-3, a typical cooling tower in a medium-size installation. The tower in larger installations is made up of several sections, each containing all of the various parts indicated in Fig. 4-3. All of the sections, also referred to as cells, are interconnected by equalizer lines to prevent water overflow from any one section.

Influencing Factors

Evaporation — When the condensing water passes over the cooling tower, a portion is lost by

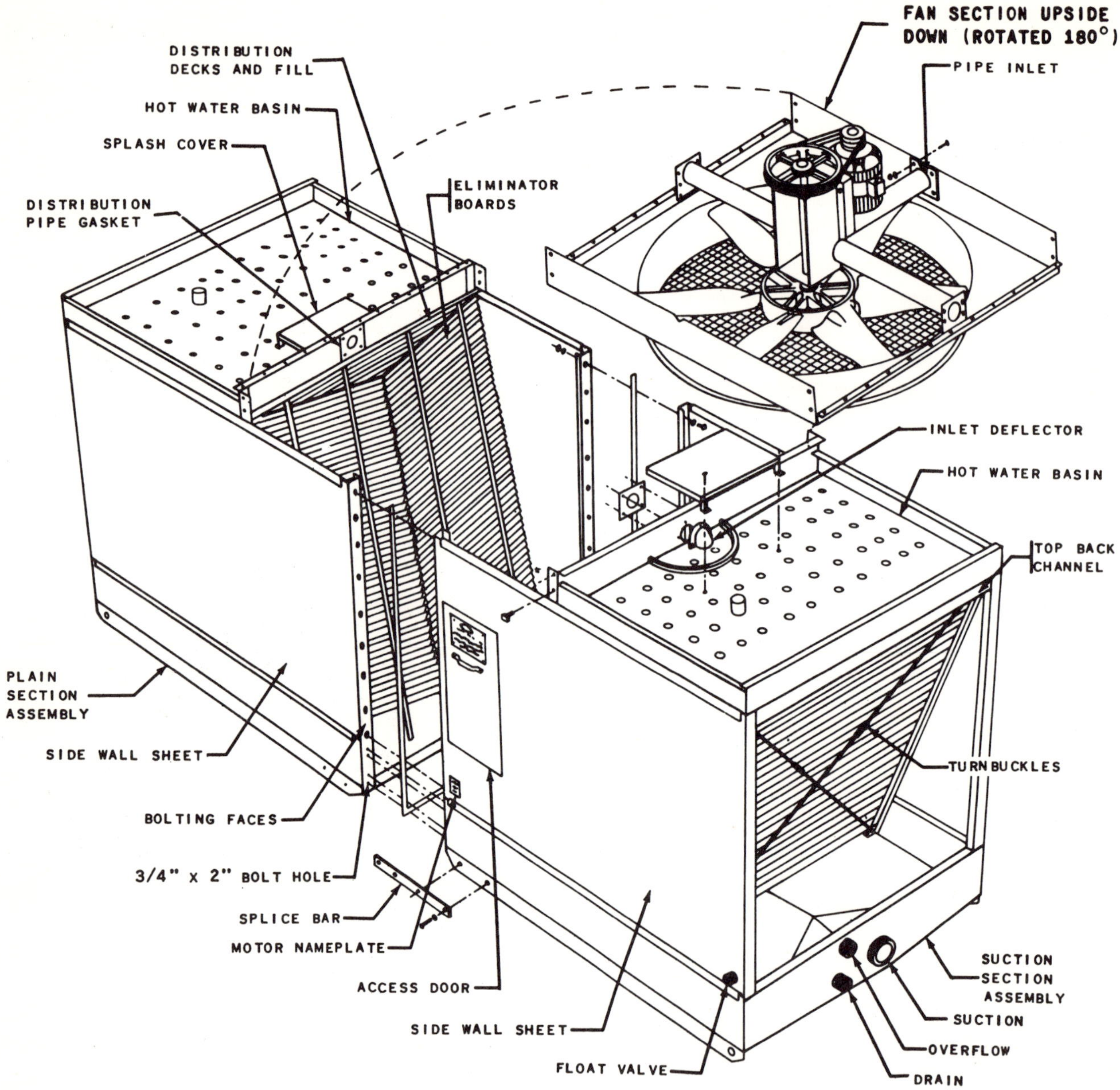

Courtesy of The Marley Company

Fig. 4-3. Water-cooling tower assembly. (For medium-size installation.)

evaporation. This loss is about 1.0 per cent of the water quantity, per 10 degree drop in water temperature, circulated through the tower. The evaporation produces the cooling effect needed to lower the temperature of the water being circulated back to the condenser. As evaporation takes place, the dissolved solids in the evaporated water are left behind, resulting in an increase in the amount of solids in the circulated water. Depending on conditions, the condensing water may become concentrated to a point of saturation so that the solids are precipitated out of the water solution as sludge or scale. Due to the physical or chemical characteristics of these solids the precipitation is usually greatest in the condenser, which is the hottest part of the system. The ability of calcium carbonate to remain in solution, for example, decreases as the temperature increases.

Air Exposure — Most cooling towers are of the mechanical draft type in which a fan forces air through the tower and over its wetted portions. Along with the desired effect of lowering the temperature of the water, the tower has an undesirable side-effect of washing out of the air any smoke, dust, dirt, bacteria, gases, etc., which may be present. The dust and dirt show up as mud or silt in the cooling-tower sump. The bacteria and other organisms promote the growth of slime and algae. Acid gases (such as sulfur dioxide from a nearby chimney) may change a neutral water to one which is quite corrosive. Mud, silt, etc., may be washed into the circulating water system and "plate out" on the condenser, as sludge.

Bleed-off — The concentration of solids may be held in check by bleeding off some of the concentrated circulating water and replacing it with make-up water which is more dilute. For this reason, it is standard practice to provide a bleed line at the tower so that a portion of the water may be deliberately wasted from the system to avoid excessive build-up of dissolved solids. Considering the purpose and importance of bleed, it is essential that every tower be equipped with a bleed line and that an adequate rate of bleed be established. The required amount of bleed is determined by evaluation of the chemical characteristics of the water, the type of water treatment, if any, to be employed, and other factors. This will be covered under Water Treatment in Chapter 5. The bleed line at the tower should be tapped into the circulating system just ahead of the inlet discharge so that water is bled to waste only when the circulating pump is in operation. The bleed line discharge should be piped to the nearest sanitary sewer.

Water Losses and Make-Up — When a cooling tower is operating, the float valve in the basin or sump of the tower admits water to the system to make up for the water that is lost. Water loss includes that lost by evaporation, by windage and splash, etc., and the loss by deliberate bleed. Disregarding splash and leakage, the amount of make-up represents the sum of evaporation, windage, and bleed. As stated previously, evaporation represents a loss of approximately 1.0 per cent of the circulating water. Windage for a mechanical draft tower represents a .1 to .3 per cent loss. Evaporation and windage are characteristics of system design. Bleed is adjusted as required.

Cycles of Concentration — The number of "concentrations" is one way of comparing the dissolved solids in the circulating water with those in the make-up or raw water. If, as a result of evaporation, the solids in the tower become twice as great as those in the make-up, there are said to be two cycles of concentration. Three times as many solids would result in three cycles of concentration; four times would be four cycles; and so on. The maximum allowable cycles of concentration is predetermined by analysis of the make-up water and other factors. After a plant is placed in operation, increasing the bleed-rate will lower the cycles of concentration; reducing the bleed-rate will increase the cycles.

Cooling-tower Problems

Spill-Over — The float should be portioned so that there is no "spill-over" into the overflow or the basin itself, when the tower is operating. This needs periodic checking to assure against wasting both water and the chemicals used in its treatment.

Water Change — The water in the circulating system is affected by: (1) the natural impurities in the make-up water, (2) the concentration of solids resulting from evaporation, and (3) the presence of air-borne particles washed out of the air as the water passes over the tower. To combat the change in water, chemicals are required. (See Chapter 5, Water Treatment.)

Wood Decay — Along with other problems, there may be encountered conditions in which the wood in the cooling tower becomes decayed or destroyed as a result of chemical or biological attack, or by rupture of the wood cells from salt crystallization. Chemical attack results in delignification of the wood in which the cellulose material in the lumber is leached out, leaving long stringy fibers of greatly reduced strength. Excessive use of chlorine seems to speed-up the process. Biological attack includes the action by fungi which destroy the wood both on the surface and from within. The fungi are reproduced from spores washed out of the air by the tower. Rupture of the wood fibers of the tower by the crystallization of salts is caused by the intermittent splashing and drying of the wooden parts, producing concentrated salt deposits through evaporation.

Evaporative Condenser

Evaporative condensers are sometimes confused with a cooling tower since they utilize a fan and water, operating within a housing similar to that of a cooling tower. Figure 4-4 shows a typical evaporative condenser. It will be noted that the condenser coil is an integral part of the unit, whereas in a cooling tower, the condenser is a component of the refrigeration plant as shown in Fig. 4-1.

Seasonal Start-Up

In order to insure the steady operation of the cooling tower and guard against costly shutdowns during the cooling season, the following tasks should be performed:

1. Brush out tower. Flush out dirt and sump.
2. Where there are wood towers, caulk cracks and openings in tongue-and-groove joints, using a soft mastic. Tighten all bolts. Paint any spots that have appeared since the shut-down painting.
3. Clean and lubricate fan bearings.
4. Install V-belts that were removed on shut-down. Adjust to proper tension and alignment.
5. Clean and lubricate electric-motor bearings per manufacturer's recommendations.
6. Operate fan-motor and fan. Check motor and fan bearing for over-heating. Also check for proper rotation and RPM.
7. Introduce chemicals for water treatment in the condenser and chiller. Where these have an automatic feeder, adjust for proper feed.
8. Fill system to proper basin level for the particular equipment. Check water-level and float adjustment.
9. Start-up cooling system and adjust bleed-rate per instructions herein, or by following supplier's recommendation.

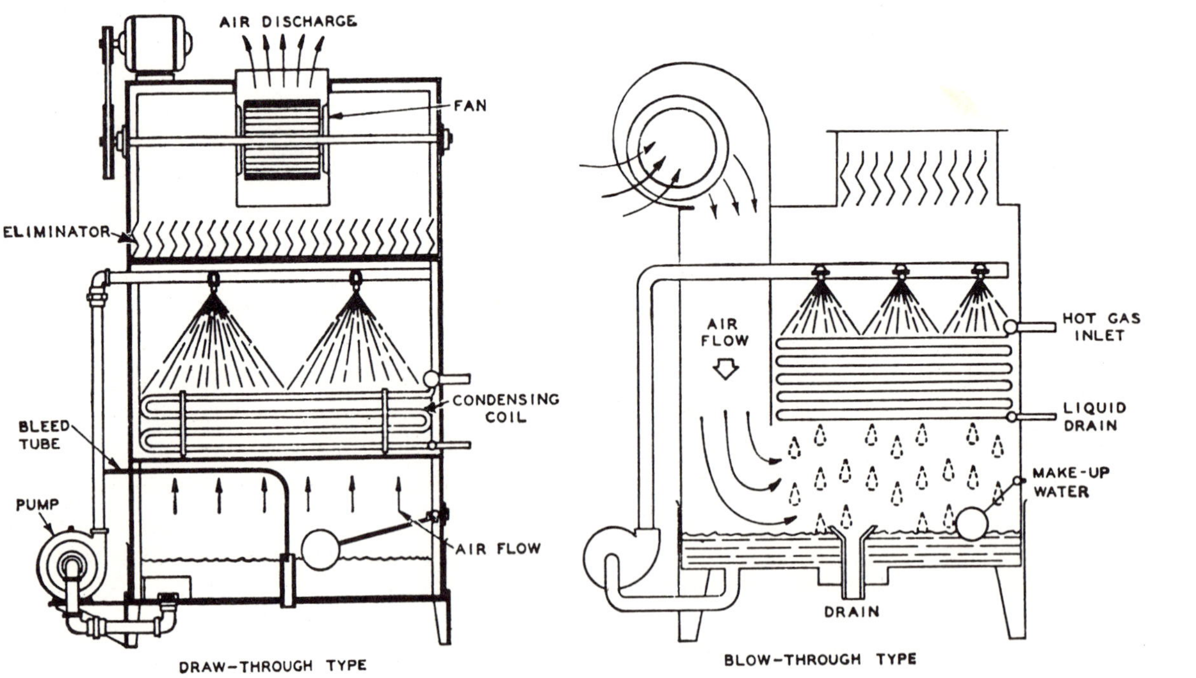

Reprinted by permission from the ASHRAE Guide and Data Book, 1965

Fig. 4-4. Evaporative condenser.

10. Make certain that spray nozzles or water distribution holes are clean and operate freely.

Operating Checks

During the cooling season the checks listed below should be made, following manufacturer's recommendations:

1. Check fan by visual inspection of V-belt conditions and alignment. Check fan and motor for unusual noise and vibration.
2. Check water-basin level and float.
3. Check bleed-rate. Adjust as required. Remember, the bleed-rate should be higher during height of the cooling season than at either the beginning or the end.
4. Observe nozzle spray (and water distribution holes) for possible clogging.
5. Check water-treatment effect.
6. Clean intake strainer.
7. Check float valve. Make adjustment for proper water-level and overflow control when required.
8. Check tower basin for accumulation of mud, silt, slime, etc. Drain, hose-out, and fill as required.
9. Lubricate fan bearings as required.
10. Check shaft couplings, if so equipped, and oil-level in gear unit.
11. Check steel casing and metal parts of tower for rust and corrosion.
12. Perform any repairs which are urgent and schedule remainder to be completed when tower is shut down at end of season.

Seasonal Shut-Down

The frequency for painting shut-down equipment may be once in two years, depending upon local conditions. Towers constructed of redwood, with corrugated asbestos casing, will require no painting.

1. Clean and flush-out tower. Leave drain open. Acid-clean, if necessary.
2. Drain only those parts of tower system likely to freeze. If necessary to drain entire system, leave all valves open. *Clean:* drift eliminators, fill-decks, louvers, spray heads, and water-distribution holes.
3. Remove V-belts from fan assembly and store out of the weather.
4. Have all required repairs made during off-season for placing tower in operation when needed. This includes leaking pump-seals and sump leaks.
5. Note any signs of rusting, and parts in need of painting.
6. When tower is completely dry, clean all surfaces that need painting. Apply a primer having a chromate base. Then apply either a marine-type paint or one recommended by the manufacturer of your tower.

Trouble-Shooting Guide

Table 4-1 is a general trouble-shooting guide for a typical cooling-tower, and should be helpful where such a guide has not been furnished by the manufacturer.

Cooling-Tower Terms

The following are some terms pertaining to cooling towers:

Cooling Range — The number of degrees through which the water in the tower is cooled. (See Fig. 4-5.)

Approach — Difference in degrees between the temperature of cold water leaving the cooling tower and the wet-bulb temperature of the surounding air. (See Fig. 4-5.)

Heat Load — The amount of heat in Btu per hour (or in Btu per minute) given up by the water as it passes through the cooling tower. It is equal to the pounds of water circulating through the tower, multiplied by the Cooling Range.

Cooling-Tower Pump Head — The amount of pressure required to lift hot water returning to the top of the tower from a point level with the base of the tower, and then forcing it through the distribution system.

Drift (Windage. See Water Treatment) — The amount of water lost to the air blowing through the tower. This water is retained

Table 4-1. Troubleshooting Guide

TROUBLE	CAUSE	REMEDY
Excessive Water Drift	Faulty drift elimination	1. Check to be sure all eliminator slats are in place. 2. Top deck assemblies and wood fill must be intact and level. There must be no tendency to channel water. 3. There must be no leaks under basin sides.
	Overpumping	Reduce water flow to tower to design conditions.
Cold Water Too Warm	Overpumping	Reduce water flow to tower to design conditions.
	Wood fill not level	Check wood fill with carpenter's level transversely and longitudinally. Level if required.
	Wood fill not properly installed	Alternate decks should have been turned end for end.
	Not enough air	Check motor current and voltage to be sure of correct contract Hp. Clean algae or muck from wood fill and eliminators.
Unusual Motor Noise	Motor running single phase	Stop motor and attempt to start it. Motor will not start if single phased. Check wiring, controls, and motor.
	Electrical unbalance	Check voltages and currents of all three lines. Correct if required.
	Ball bearings	Check lubrication. Replace bad bearings.
Motor Runs Hot	Motor overload, wrong voltage or unbalanced voltage	Check voltage and current of all three lines against nameplate values. Repitch fan blades if necessary.
	Bearings overgreased	Remove grease reliefs. Run motor up to speed to purge excessive grease.
	Excessive belt tension	Reduce tension and check shaft alignment.
	Poor ventilation	Clean motor and check ventilation openings.
	Wrong grease	Change to proper grease. See motor manufacturer's lubricating instructions.
	Winding fault	Check with ohmmeter.
Fan Noise	Loose bolts in fan guard	Check and tighten if necessary.
Unusual Fan-Drive Vibration	Loose bolts and cap screws	Tighten all bolts and cap screws on all mechanical equipment.
	Unbalanced fan	Be sure blades are properly positioned in correct sockets. (See match numbers.) All blades must be pitched the same.
	Unbalanced motor	Disconnect load and operate motor. If motor still vibrates, rebalance rotor.
	Worn fan-shaft bearings	Check bearings and mounting bolts.
	Bent shaft	Make sure fan and motor shafts are straight and properly aligned.
	Belts	Belts should be stretched to the proper tension.

Courtesy of The Marley Company

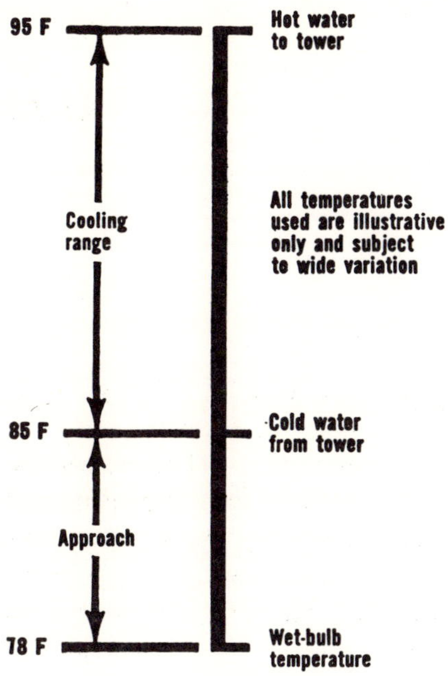

Courtesy of The Marley Company

Fig. 4-5. Diagram illustrating "Cooling Range" and "Approach."

in the air as opposed to that lost to evaporation.

Bleed — The continuous running-off to waste of a certain amount of the circulating water system, planned to lower or prevent the build-up of solids in the water. This is covered more fully in Chapter 4, under Water Treatment.

Make-Up — The water required to replace that lost to evaporation, windage (or drift), and bleed.

Cooling Tower Data — A cooling tower must meet the design conditions of a particular cooling load. To do this, certain data must be known for selecting the proper type and size of tower. This data is as follows:

a. Design for Wet-bulb temperature
b. Temperature of hot water to the tower
c. Temperature of cold water leaving the tower
d. Gallons of water per minute (GPM) to be circulated over the cooling tower
e. If mechanical draft, such electrical characteristics as voltage, phase, and frequency must be known.

Water-cooled versus Air-cooled Equipment

As stated previously, the growing need for conservation of water has increased the effort of manufacturers to produce air-cooled equipment that will handle larger cooling loads without the use of evaporative condensers or cooling towers. At present, equipment is available with the Compressor-Condenser section mounted remotely from the Fan Evaporator section (Split System), utilizing air-cooled condensers capable of providing up to 100 tons of refrigeration and, through special adaptations, possibly more. The use of these air-cooled condensers not only eliminates the need for water treatment and reduces water costs, but also permits operation in freezing weather without the modifications and extra attention required by evaporative condensers and cooling towers. Although this sounds like the complete answer, it has its limitations; for one thing, a refrigerant line which is too long is apt to develop leaks. The question of deciding which is the best method for a particular building is not an easy one. It requires an economic study based on initial cost, investment cost, depreciation, fuel and utility costs, repair and maintenance costs. This can become quite involved and requires the services of a qualified engineer to determine the most advantageous equipment and system for a particular installation.

CHAPTER 5

Water Treatment

General

The goal of a water treatment program is to provide suitable preventive maintenance at reasonable cost to preserve the equipment, maintain its efficiency, prolong its life, and avoid malfunction or breakdown. Administering the program requires an appreciation of the problems which may arise, a rudimentary knowledge of the types of chemicals commonly used, an understanding of the various methods for introducing the chemicals into the water systems, and a concept of the amount of control to be maintained on the dosages.

Biological Fouling — Biological fouling in towers is usually first recognized by the sudden growth of algae and slime in the cooling tower and the sump. Along with the algae and slime there may be present various kinds of bacteria which interact with iron and sulfur to cause corrosion and produce insoluble salts. Algae consists of tiny plant cells which multiply and produce large masses of plant material in a short time. The growth is usually most rapid in hot weather and in bright sunlight. For this reason, some operators have found it helpful, in hot weather, to place a plywood cover or umbrella over the tower to shield it from the sun. Slime growths consist of a gelatinous mass that clings to practically all surfaces in the system, where it traps organic matter, debris, and scale-forming material. When the micro-biological growths break loose and slough off the tower, they may easily clog lines, pumps, and heat exchangers. The growth also attacks and destroys the members of redwood towers. The key to combating algae and slime is mechanical cleaning followed by chemical treatment. Good housekeeping requires periodic cleaning of the tower and sump to remove mud, silt, leaves, twigs, algae, slime, etc., which may have been collected. (This will be more fully covered in Chapter 8, Preventive Maintenance.)

Scale and Corrosion — The evaporative process used to cool water in the condensing-water circulating system results in the concentration of dissolved solids in the water. Scale-forming units are produced, which are deposited on surfaces in contact with the water. Deposits of calcium carbonates from the condensing water are normally the chief scale problem, although deposits of calcium and magnesium silicates and calcium sulfate may also be encountered. Since the solubility of calcium carbonate decreases as the temperature increases, the deposits usually show up first in the tubes of the condenser which is the hottest part of the system. A build-up of scale in the condenser lowers the rate of heat transfer, reduces the efficiency of the refrigeration machine, results in an increase in consumption of electricity and may cause the refrigeration machine to cut-out on high head pressure. In some water (such as those fairly low in hardness) the deposit of scale can be minimized by adequate bleed at the tower. In others, bleed, together with chemical treatment, is required. Corrosion is generally due to the presence in the water of dissolved oxygen, acid gases, etc. The chief offenders among the acid gases are carbon dioxide and sulfur dioxide picked up from the atmosphere.

Cycles of Concentration — The number of "concentrations" is one way of comparing the dissolved solids in the circulating-water with those in the make-up, or raw water. If, as a result of evaporation, the solids in the tower water become twice as great as those in the make-up or raw water, there are said to be two cycles of concentration. Three times as much solids would result in three cycles of concentration; four times as much would be four cycles, and so on. The maximum allowable number of cycles is predetermined by analysis of the make-up water and other factors.

Bleed-Off — The concentration of build-up of solids in the cooling-water circulating system is controlled by deliberately bleeding off a certain amount of water from the circulating-water system. The float in the cooling-tower basin compensates for this deliberate bleed-off by automatically letting in raw or make-up water while also taking care of the water lost by evaporation and windage. Increasing the bleed-rate will *lower* the cycles of concentration, while reducing the bleed-rate will *increase* the cycles of concentration.

pH Value (Acidity versus Alkalinity)

Measurement — Water is either acid, alkaline, or neutral, depending on its constituents. A neutral water is one which is neither acid nor alkaline. The control of acidity or alkalinity is an important aspect of the water treatment program. The principal measurement for acidity or alkalinity is the pH of the water.

pH Value — In practical application, the pH of a water is a number between 0 and 14 which indicates the degree of acidity or alkalinity of the water. The mid-point of the scale is taken as being 7.0. A solution with this pH is considered neutral. Solutions with pH values below 7.0 are considered acid and those with values above 7.0 are considered alkaline. A pH below 7.0 encourages corrosion of equipment; a pH above 7.5 encourages the deposit of calcium carbonate scale. Normal pH values are:

a. *Chilled-water system* — Normally it is desirable to maintain a pH of about 7.5 to 8.0 in the chilled-water (or closed) system.
b. *Cooling-tower system* — Normally it is desirable to maintain a pH of about 7.0 to 8.0 in the cooling-tower (or open) system.

Change in pH — Where city water is used as make-up, the pH of the incoming water is usually in the range of 6.0 to 8.0. Once the water is circulated in the system, the pH will almost always change because of the water treatment required, cycles of concentration, foreign material picked up in the system, etc. Thus, the pH will vary with the installation and the kind and extent of the water treatment required.

Measuring pH — There are several ways of measuring pH. For the usual installation where no local condition justifies using more expensive equipment, color indicators may be used. In this method, the color of the water itself or specially-treated paper (litmus) inserted in the water is compared with standard shades of colors over the range of a particular indicator.

Hardness — Hardness of water may be recognized in everyday life by the difficulty in obtaining lather without using a good deal of soap and by the formation of scale in vessels in which the water is boiled — in a tea kettle — for example. To evaluate the role of hardness in the operation of a cooling tower, it is sufficient to note that hardness is a quality shown by water due to certain substances in solution and that the two principal constituents causing it are calcium and magnesium. Since hardness data is normally readily available from the local water company, it is convenient to relate the program of water treatment (particularly for cooling-tower systems) to the hardness of the local water. This we have done in setting-up bleed rates, in table and chart forms, for the smaller and the larger installations, respectively.

Program — The water-treatment program, as applied herein, is separated into two groups: the chilled-water system, and the condensing-water system.

Chilled-Water System

Installations Under 100 Tons

a. *General* — We have seen in Fig. 3-8, how water, chilled by the refrigeration machine, is circulated through the chilled water coil(s) in air-handling units. The air blowing over the coil is cooled and thus provides the cooling needed to lower the temperature of the air supplied to the occupied space. On leaving the coil, the water is circulated back to the refrigeration machine and is used over and over again. A chilled-water system may be classed as a "closed system" since it is not exposed to atmosphere, except through the over-flow opening at the surge tank.

b. *Problems* — At the end of the cooling season, draining of those parts of the system which may be exposed to freezing temperatures is required. Draining the system completely each season is not recommended. Corrosion is the problem most frequently encountered in chilled-water systems. To combat corrosion, a chemical inhibitor is usually introduced into the system initially.

c. *Water Treatment* — A chemical commonly used as an inhibitor is sodium chromate used together with an added material such as caustic soda (commonly known as lye), to keep the water slightly alkaline in the range of 7.5 to 8.0 pH. Treatment is essentially a *one-time* operation to bring the concentration of inhibitor up to a desired level. Replenishment of small amounts of the treatment may be needed from year to year because of loss of water through pump glands, possible overflow at the expansion tank, and drainage of parts of the system likely to freeze.

d. *Dosages* — Table 5-1 has been set up for installations under 100 tons or where a water-treatment service contract is not deemed necessary. The table is based on tons of refrigeration. Where there are two refrigeration machines and no common header in the chilled-water system where the dosage can be introduced, place one-half the dosage into each chiller; if there are three machines (and no common header), place one-third the dosage in each chiller, etc. Depending on conditions, it may develop that more complex treatment may be needed than that prescribed in the table. In this case, a water-treatment service contract should be obtained. (See Chapter 10, Service Contracts.)

e. *Introducing Chemicals into the System*

(1) *By-Pass Feeder* — Where a by-pass feeder has been provided, the chemical pot is first valved off and loaded with the desired chemicals, either in dry form or in solution. The valves are then adjusted so that the water flushes the chemicals into the system.

(2) *No By-Pass Feeder* — Where a by-pass feeder has not been provided, it will be necessary to develop some other means for introducing the chemicals. On some jobs it may be possible to load the baskets in the line-strainers with the chemicals and let the water flush the chemicals through them. On others, a means may be devised of sucking a chemical solution into the chilled-water line from the suction side of the pump or forcing it into the lines with a special pressure pump. On some jobs the chemicals have been fed into the expansion tank after some of the water in the system has been drained off.

Table 5-1. For Installations Under 100 Tons (Chilled-Water System)

Chemical	Tons of Refrigeration						
	30	40	50	60	70	80	90
	Pounds of Chemical						
Sodium Chromate	2.0	2.5	3.0	3.5	4.0	4.5	5.0
Caustic Soda (Lye)	1.0	1.25	1.5	1.75	2.0	2.25	2.5

Condensing Water System

General — We have seen in Fig. 4-1 how the water from the tower circulates through the condenser and back to the tower. In contrast to the chilled-water system, which is classed as a "closed system," the condensing water from the tower is classed as an "open system." The fact that it is an open system makes it subject to various adverse atmospheric conditions. Thus, there is need for more treatment than for the water in the closed or chilled-water system. In this handbook, only the matter of proper bleed will be covered in detail for both small and large installations. In regard to water treatment, it will be covered to a certain degree for installations under 100 tons. For installations of 100 tons and over, it is recommended that the services of a reputable water-treatment company be utilized. (See Chapter 10 for sample specifications.)

Installations Under 100 Tons

Bleed — The initial bleed should be established to maintain the "cycles of concentration" in line with the "hardness" of the local water supply as shown by the following relationship:

Hardness of Make-Up Water (p.p.m.)	Maximum Cycles of Concentration
Less than 60	5
60 to 120	4
121 to 180	3
More than 180	2

Based on the above relationship, the bleed shall be established in accordance with Table 5-2.

Table 5-2. Bleed-rate for Installations Under 100 Tons.*

	Hardness of Local Water (p.p.m.)				NOTE
	Under 60	60 to 120	121 to 180	Over 180	Bleed should operate *only* when cooling-tower pump operates.
	Gallons per hour				
Bleed-Rate	3	5	7	16	

* The table is based on a 10-ton system. For larger systems (up to 100 tons), adjust the bleed-rate proportionally.

Bleed Adjustment — It is desirable to maintain the cycles of concentration as high as the quality of the water will permit, while at the same time keeping the bleed-rate at the lowest possible point in order to avoid wastage of water and chemicals. As stated previously, increasing the bleed lowers the cycles; reducing the bleed increases the cycles. Assuming the correct initial bleed-rate was established at the beginning of the cooling season, it may be found necessary to increase the rate due to increased load, then decrease the rate again as the end of the cooling season approaches.

Water Treatment — As stated previously, the subject of water treatment will not be covered in detail. However, a few general guidelines for the small-to-medium size installation, relative to Scale and Algae Control, follow:

a. *Scale Control* — For scale control, chemicals in some form of phosphate are available from air-conditioning concerns or directly from suppliers. The chemical is furnished in a slow-dissolving form. It is placed in a mesh bag and hung in the tower above the level of the sump so that the contents of the bag will be wet when the tower is operating and will drain dry when the tower shuts down. If there is insufficient room to hang the bag(s) above the water level, place them under the water spray or at the top of the tower where the water comes in. Avoid placing the bags in a strong stream of water which would cause the chemical to dissolve too rapidly, or in a position where they might restrict the flow of air. Replenish the chemical as it is exhausted. For normal operation, about 20 per cent may dissolve each month. The suggested initial dosage at the beginning of the cooling season is 4 lbs. for each 10 tons of refrigeration. Remember to replenish as required.

b. *Algae Control*—For algae control, chemicals in the size of pellets or briquettes, of some form of chlorophenate are available from air-conditioning concerns or directly from the suppliers. Whenever growths of algae appear, place the prescribed amount of chlorophenate in the sump of the cooling tower. (Usually about 1 lb. for each 10 tons of refrigeration.) Increase the quantity if the growths are persistent. Shut off the bleed line for two or three hours during the day or overnight at the time the chlorophenate is put into the tower. When the algae become dislodged, drain the sump and readjust the bleed-rate to its original setting.

ESTABLISHING INITIAL BLEED

INSTALLATIONS 100 TONS AND OVER

1. SUGGESTED CYCLES OF CONCENTRATION: (Relationship Table)

Hardness of local Water in P.P.M.	MAXIMUM CYCLES OF CONCENTRATION
less than 60	5
61 to 120	4
121 to 180	3
More than 180	2

2. BASIC CONSIDERATIONS:

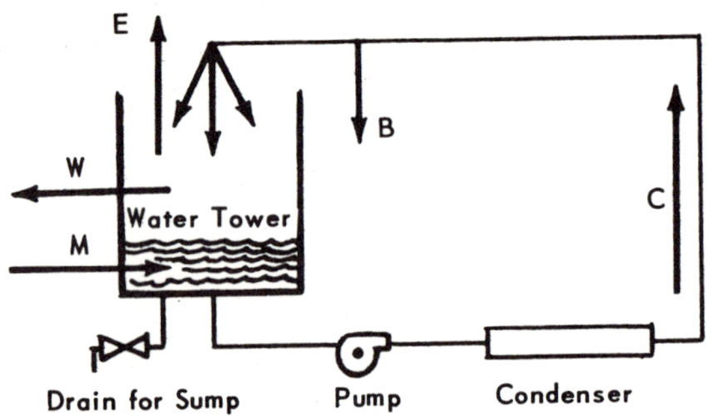

C = Circulation Rate, 3 g.p.m. per ton
E = Evaporation, 1% C
W = Windage 0.1% to 0.3% C
B = Bleed
M = Make-up = E + W + B

3. BLEED-RATE IN RELATION TO TONNAGE AND CYCLES OF CONCENTRATION:

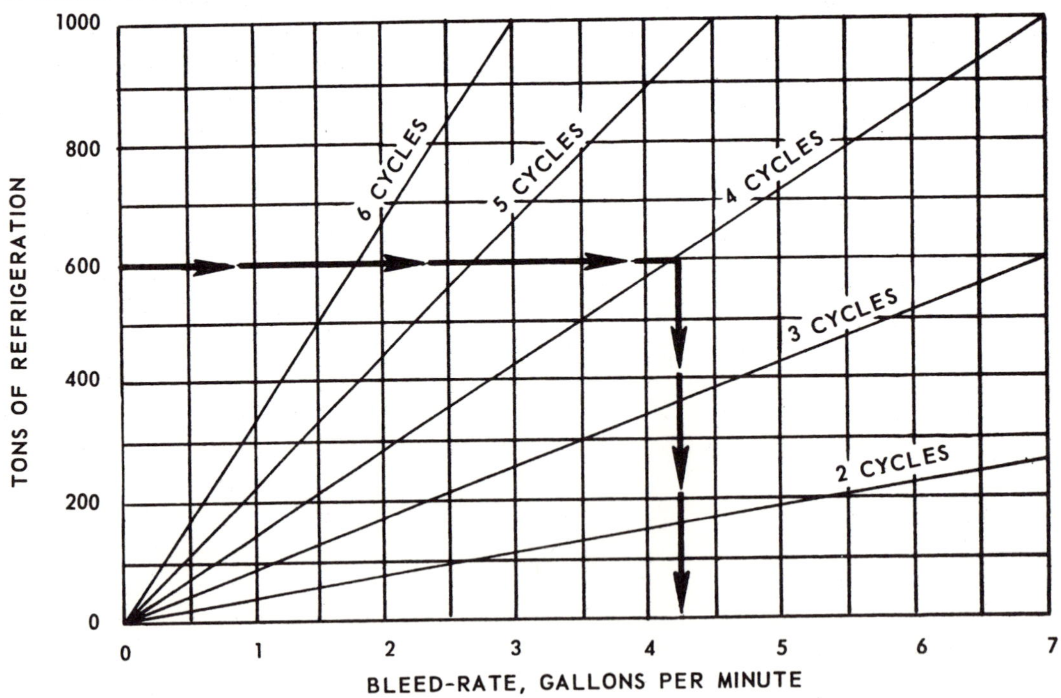

4. EXAMPLE: 600 Tons; Hardness 100 p.p.m.; 4 Cycles; Circulation 1800 g.p.m.; Bleed 4-1/4 g.p.m.

5. NOTE: For absorption systems, use twice as much bleed as indicated above.

Fig. 5-1. Establishing initial bleed. (For installations of 100 tons and over.)

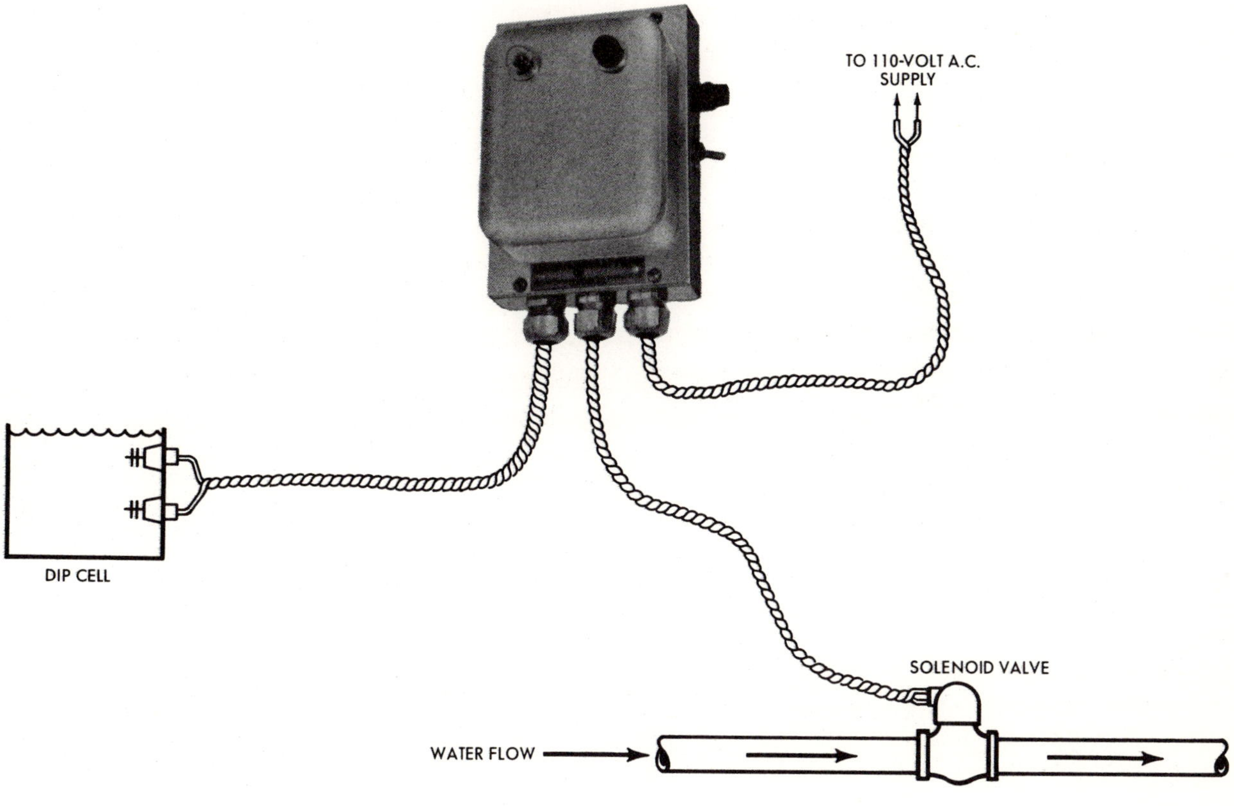

Courtesy of North American Mogul Products Co.

Fig. 5-2. Automatic bleed-control device.

Sodium hypochlorite, commonly known as laundry bleach, and available in 5 per cent liquid solution from supermarkets or grocery stores, may be used when it is found that the chlorophenate is proving ineffective after several feedings. Alternating the chemicals is usually found to be helpful. Suggested dosage: 2 lbs. for each 10 tons of refrigeration.

c. *Corrosion Control*—Corrosion control requires a daily check to make certain the pH does not drop below 7.0. Should this occur, the services of a water-treatment concern should be engaged.

CAUTION: It should be remembered that the above dosages are approximate and are only generally suggested. It would be foolish to think that water treatment was as simple as this. Raw make-up water differs greatly in different localities and also, the requirements of individual installations will differ. Where unusual water conditions exist or where much trouble is experienced, the services of a professional water-treatment concern should be utilized.

Installations 100 Tons and Over

Bleed — The same procedure for setting up the initial bleed for the smaller installations should be followed in setting up the bleed for the larger installations except that, instead of using the bleed table, the bleed-rate should be obtained from the bleed chart. (See Fig. 5-1.)

Bleed Adjustment — The same comments and procedure outlined for the small installation also apply to the larger installations.

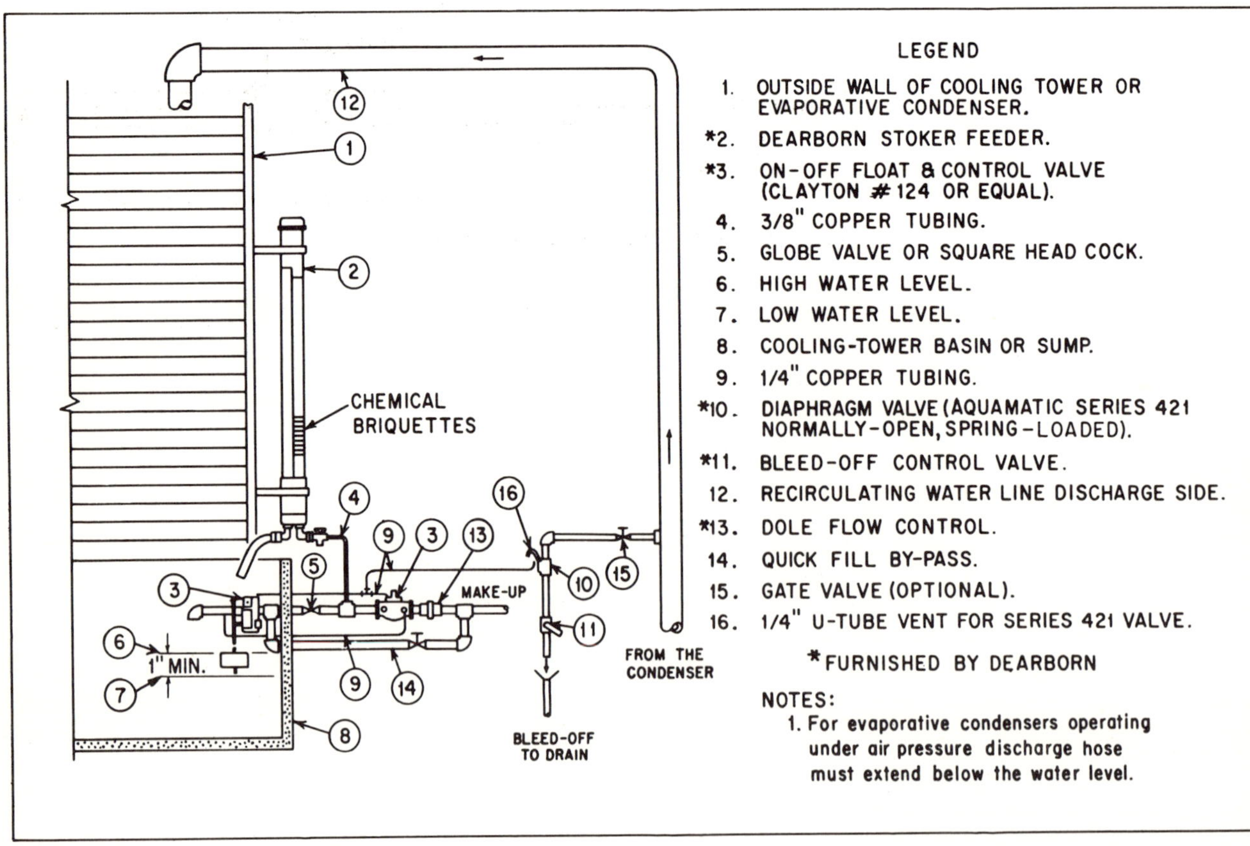

Courtesy of Dearborn Chemical Div., W. R. Grace & Co.

Fig. 5-3. Automatic feed and bleed-off system for briquette chemicals.

Automatic Control Equipment (For Bleed and Chemicals)

There are on the market, relatively simple devices at nominal cost, that automatically control (1) bleed only, (2) chemicals only, or (3) both bleed and chemicals. The bleed controls are effective in maintaining a satisfactory concentration of solids in the condensing-water system and in reducing the amount of make-up water required.

Fig. 5-2 shows a device available for control of bleed only.

Fig. 5-3 shows a device that controls both bleed and chemicals. This briquette-type feeder is usually limited to installations of up to 300 tons of refrigeration. The briquette-type feeder is also available for feeding chemicals only.

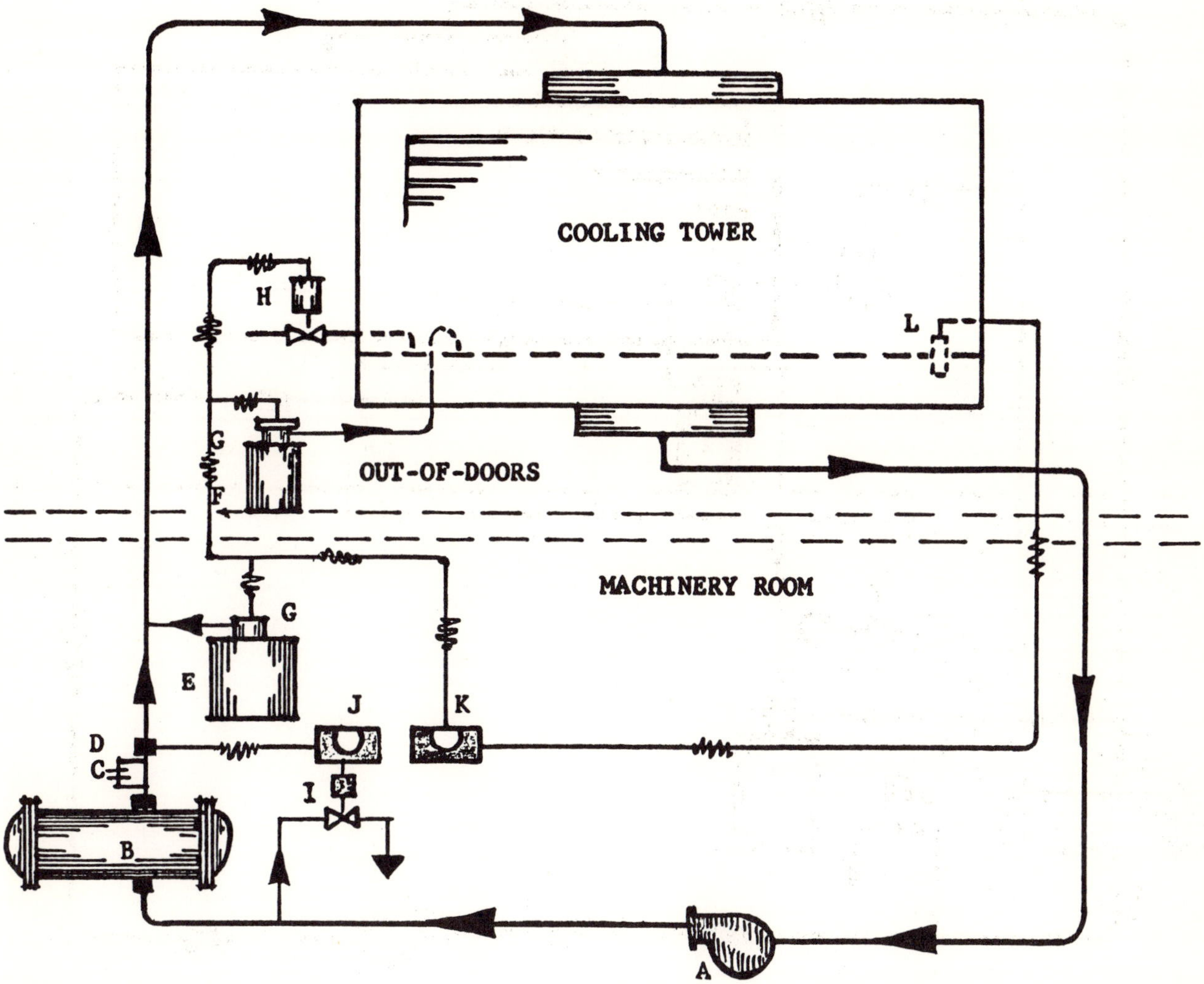

(A) Pump; (B) Condenser; (C) By-pass nipple for inspection; (D) Blow-down sensing probe; (E) Chemical tank; (F) Acid tank; (G) Adjustable pump; (H) Solenoid make-up valve; (I) Blow-down solenoid valves (1 or more); (J) Blow-down relay; (K) Master controller; and (L) Level-sensing probe.

1. Water level actuates control.
2. Chemical- and acid-proportioning pumps, adjustable for quantity of water through orifice make-up. (Acid-feed equipment only where determined necessary by study of water analysis.)
3. Blow-down controlled by conductivity.
4. Both pumps and blow-down valve should be capable of manual operation.

Fig. 5-4. Layout #1—Automatic feed of water-treatment chemicals. (For large installations.)

Figs. 5-4 and 5-5 are suggested piping layouts for automatic-feed, water-treatment chemicals for the larger installations.

a. Layout No. 1 — The water level in the tower activates the control.
b. Layout No. 2 — Impulses from the make-up-water meter are accumulated to activate the timer assembly when a pre-set quantity of water has passed through the meter.

Make-up Water — The conditions shown to the right of the tower diagram on the bleed chart (Fig. 5-1) are the average conditions for a mechanical-draft tower having a water-temperature-drop design of 10 degrees. The percentages shown give

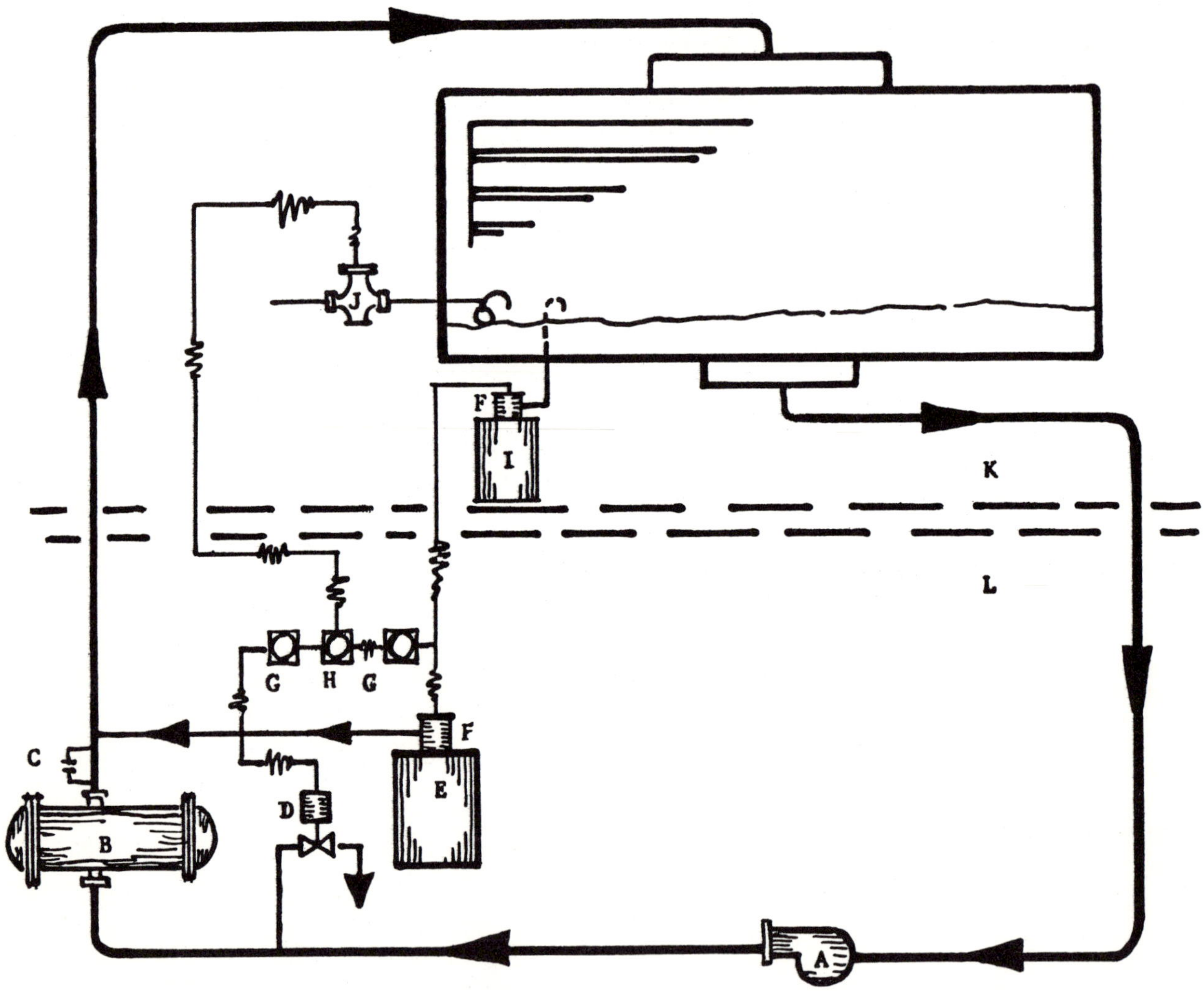

(A) Pump; (B) Condenser; (C) By-pass nipple for inspection; (D) Motor blow-down valve; (E) Chemical tank; (F) Adjustable pump; (G) Timer; (H) Impulse accumulator; (I) Acid tank; (J) Contact register make-up-water meter; (K) Out-doors; and (L) Machine room.

1. Impulses from make-up-water meter are accumulated to actuate timer assembly when pre-set quantity of water has passed through meter.
2. Timer regulates operating time of acid and chemical pumps and motor-operated blow-down valve for each cycle.
3. Adjustable proportioning pumps feed proper quantities of acid and chemicals for each cycle of operation. (Acid-feed equipment only where determined necessary by study of water analysis.)
4. Both pumps and blow-down valve should be capable of manual operation.

Fig. 5-5. Layout #2—Automatic feed of water-treatment chemicals. (For large installations.)

an idea of the large amount of make-up water required due to loss through evaporation, windage, and bleed. Using the figures from the example at the bottom of the bleed chart (Item # 4), the make-up water required when the tower pump is operating would be as follows:

Loss Due to Evaporation:	.01 x 1800 =	18.0 g.p.m.
Loss Due to Windage:	.002 x 1800 =	3.6 g.p.m.
Loss Due to Bleed:	... ...	4.25 g.p.m.
	Total:	25.85 g.p.m.

From this average condition, it can be seen that bleed alone accounts for approximately 16 per cent of the total make-up water. Proper control over this item can and should be exercised. Under normal conditions, very little control is possible over evaporation and windage.

Water Treatment Costs

The many chemical compounds in use by water-treatment companies all vary greatly in composition and cost depending on the purpose for which they are compounded. The range in cost for the dry kind is as much as from 20 cents per pound (or less), to as much as 55 cents or more. To give a figure for the cost of water treatment would be meaningless since the requirements of each individul installation differ. The figures shown below are per ton of refrigeration per month in use, and will give some idea of the cost. These figures also could vary as much as 50 per cent above or below the costs shown.

Under 50 Tons:	$0.30
50 to 100 Tons:	0.25
100 to 250 Tons:	0.20
250 to 500 Tons:	0.15
500 to 1,000 Tons:	0.10

It is strongly urged that, where competent personnel are not available, the services of a reliable water-treatment concern be utilized, at least at the beginning. Proper water treatment and preventive maintenance go together.

Summary

The following statements on water and water treatment are taken from the Calgon Corporation's little booklet entitled *Cooling Water on the Job* and sum up the subject of water treatment very aptly:

> *Scale* — Scale is the solid matter left after water is heated or evaporated. It forms thick deposits on heat transfer surfaces and prevents the heat from being carried off.
>
> *Corrosion* — Corrosion is caused by acid gas and other substances dissolved in the water. Slowly but surely, corrosion eats through metal parts and ruins the equipment.
>
> *Algae and Slime* — These are microscopic living plants and organisms that grow and multiply in water. They will clog water-distribution holes, pump suction-screens and also reduce heat transfer in condensers.

Scale and slime deposits will reduce water flow, cut heat transfer, and cause high head-pressures. The compressor must work harder and draws extra current, thus increasing power costs and causing equipment to fail to produce the designed capacity. The life of the equipment itself is reduced. In fact, excessively high head-pressure is a common cause of motor burnout.

When water is warmed or evaporated and the concentration of minerals increases past the saturation point, the scale drops out and builds up on condenser surfaces, piping, water-regulating valves, and tower fill. Not only is the flow of water reduced, but the scale acts as an insulator, reducing the heat transfer.

CHAPTER 6

Air Supply Systems

Types — In this handbook, only three types of air supply systems will be referred to directly:

a. Individual Room Fan-coil Units. (See Fig. 6-1.)

b. High-velocity Induction System. (See Fig. 6-3.)

c. Central Air Handlers Serving Ceiling Diffusers. (See Figs. 6-5 and 6-6.)

All the many variations in air-distribution systems would be too large a field to cover, for the purposes of this handbook. Since the central air handler is the most difficult system to understand, this type — together with its controls — will be the only one covered here, in detail.

Individual Room Fan-coil Units

Figure 6-1 shows a typical Fan-coil Unit which is usually located beneath a window. The main components of the fan-coil unit (shown in the inset) are: (1) a finned (heating or cooling) coil, 2) a fan (usually twin), and (3) a sectional filter. As in the case of the air handler, hot or cold water is piped to the coil of the unit from a central heating or cooling plant. It may or may not provide

Courtesy of McQuay, Inc.

Fig. 6-1. Typical, individual room fan-coil unit.

ventilation; but to do so, openings are provided in the wall in back of the unit. Fig. 6-2 shows a typical Fan-coil distribution system. Note that, in this case, through-the-wall openings behind the units are provided for ventilation.

Function — A typical fan-coil unit performs in the following manner:

a. It blows air over a coil.
b. Depending upon whether cold or hot water is being circulated through the coil at the time, it cools or heats the air flowing over the coil.
c. Filters the air.
d. Temperature control is obtained through operation of the fan (normally activated by a room thermostat). The fan usually has 3 speeds: high, medium, and low, which are changed manually, except for unusual situations or conditions.

Types of Piping Systems:

a. *Two-pipe System* — Has one supply line and one return line.
b. *Three-pipe System* — Hot water and chilled water are available the year round. Depending on space needs, a control valve at the unit selects either hot water or chilled water from the secondary hot-water or chilled-water pump. There is one common return.
c. *Four-pipe System* — The primary difference between this system and the 3-pipe system is that a separate return line is furnished for *both* the secondary hot-water and chilled-water supply lines.

Advantages:

a. Requires no ducts for air distribution (except for a small amount when induction units are installed). Installation in existing buildings is much easier and much less disrupting to occupants and, in some cases, the cost may be much less.
b. Units are easy to service and replace.
c. Requires no mechanical ability on the part of maintenance personnel except in the case where 3- and 4-pipe systems are installed.
d. Operation and temperature control are not complicated.
e. No uncomfortable drafts.
f. Occupants have full control of temperature desired.
g. The 3- and 4-pipe systems can provide heating and cooling all year round.
h. Initial cost can be less than with air handlers in cases where ventilation is not provided for the fan-coil units through wall openings behind the units.

Disadvantages:

a. Humidity control is not as good as that provided by air handlers.
b. Trouble may be experienced with clogged condensate drain pipe due to slime and algae. Causes leaks on floors, damaging to costly apartment-building carpeting.
c. In large buildings, the number of units involved and the frequent inspections required to avoid possible damage from leaks, can result in a costly number of maintenance personnel man-hours expended.
d. In a 2-pipe system, changeover from heating to cooling and vice-versa becomes a problem at the beginning and end of the cooling season. Occupants become very dissatisfied and numerous complaints are received, particularly from those paying high rents.
e. Plant operator must use extreme caution when making changeover (particularly from heating to cooling), to prevent overloading the chiller with an onrush of hot water with a resultant shock and possible damage to the machine.
f. When fan-coil units are placed under windows, which is normal practice for best results, particularly in cases where ventilation is provided through outside wall openings behind the units, the protruding unit interferes with the fall of full drapes in luxury apartments.

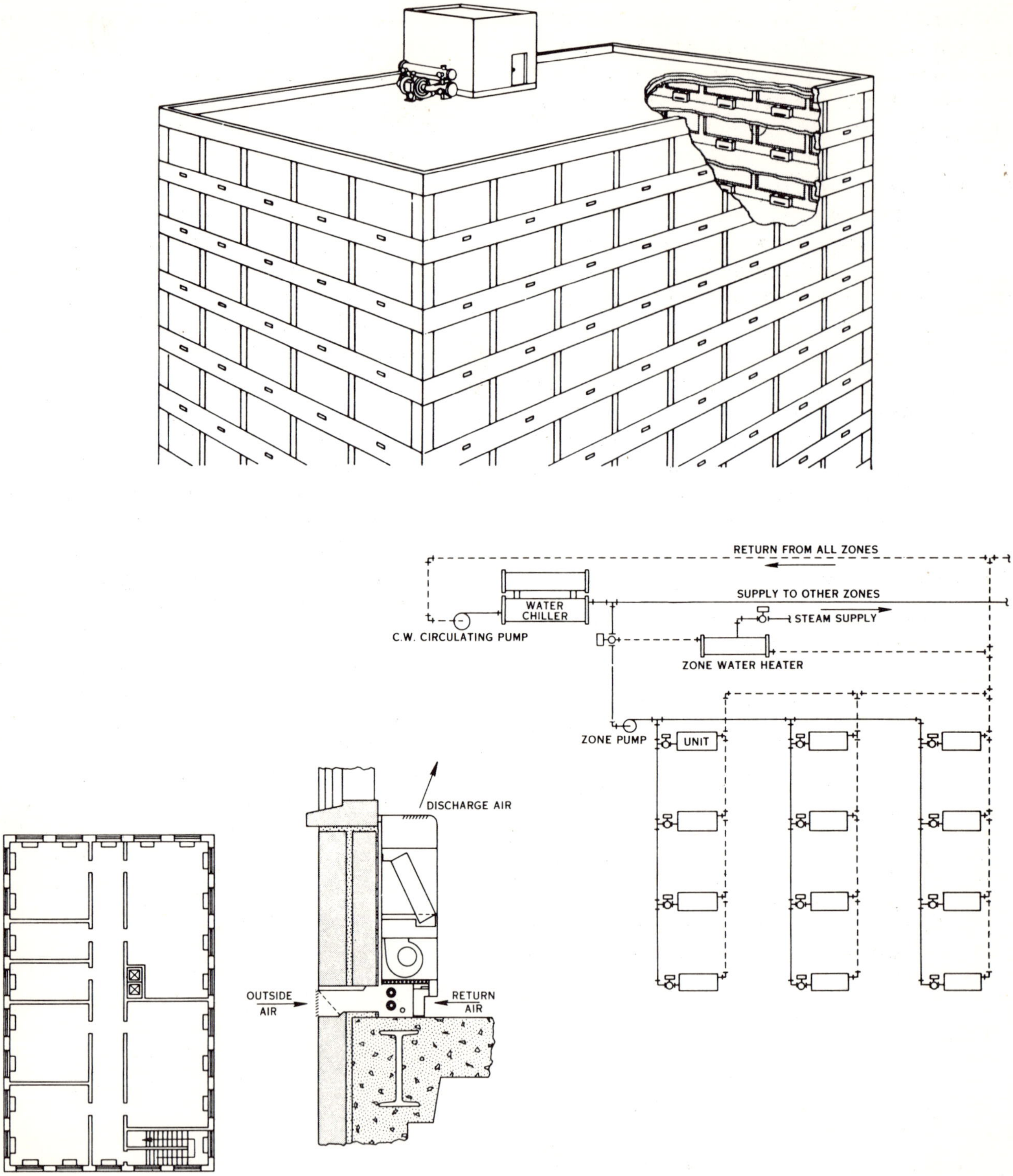

Courtesy of The Trane Company

Fig. 6-2. System with fan-coil filter units in each conditioned space. Water piping to each unit; refrigeration unit for supplying chill-water; chilled-water circulating pump; water heater for each zone; circulating pump for circulating hot or cold water in each zone; an outside air intake on each room unit; and controls.

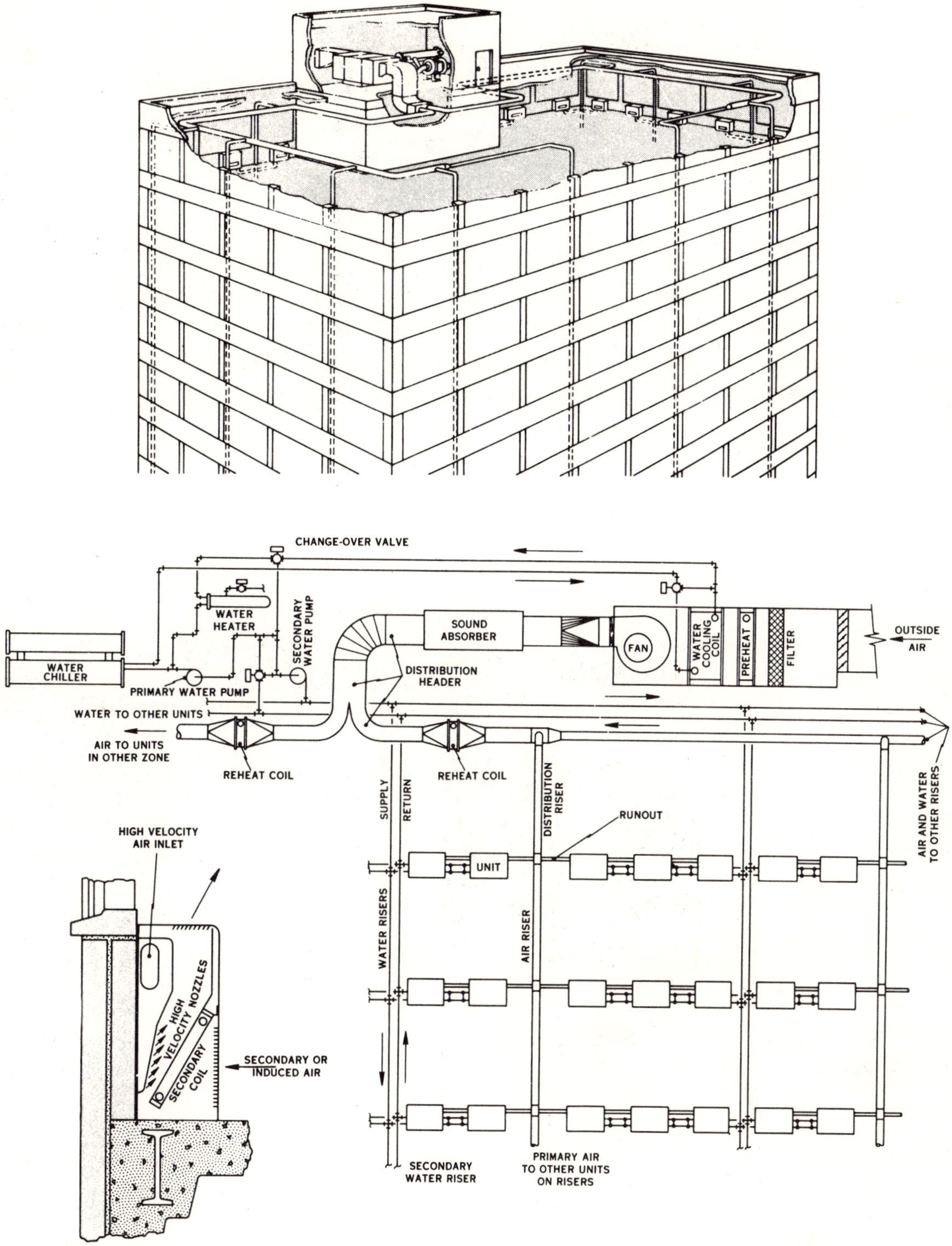

Courtesy of The Trane Company

Fig. 6-3. Central fan-unit system with cooling and dehumidifying coil; a preheat coil and filters; a refrigeration unit for chilling water; a water-circulating pump for primary water and one for secondary water; zoned high-velocity ductwork for distributing the high-velocity primary air to perimeter induction units located under the windows; a zone-reheat coil for each zone; a water heater for heating the secondary water; and controls.

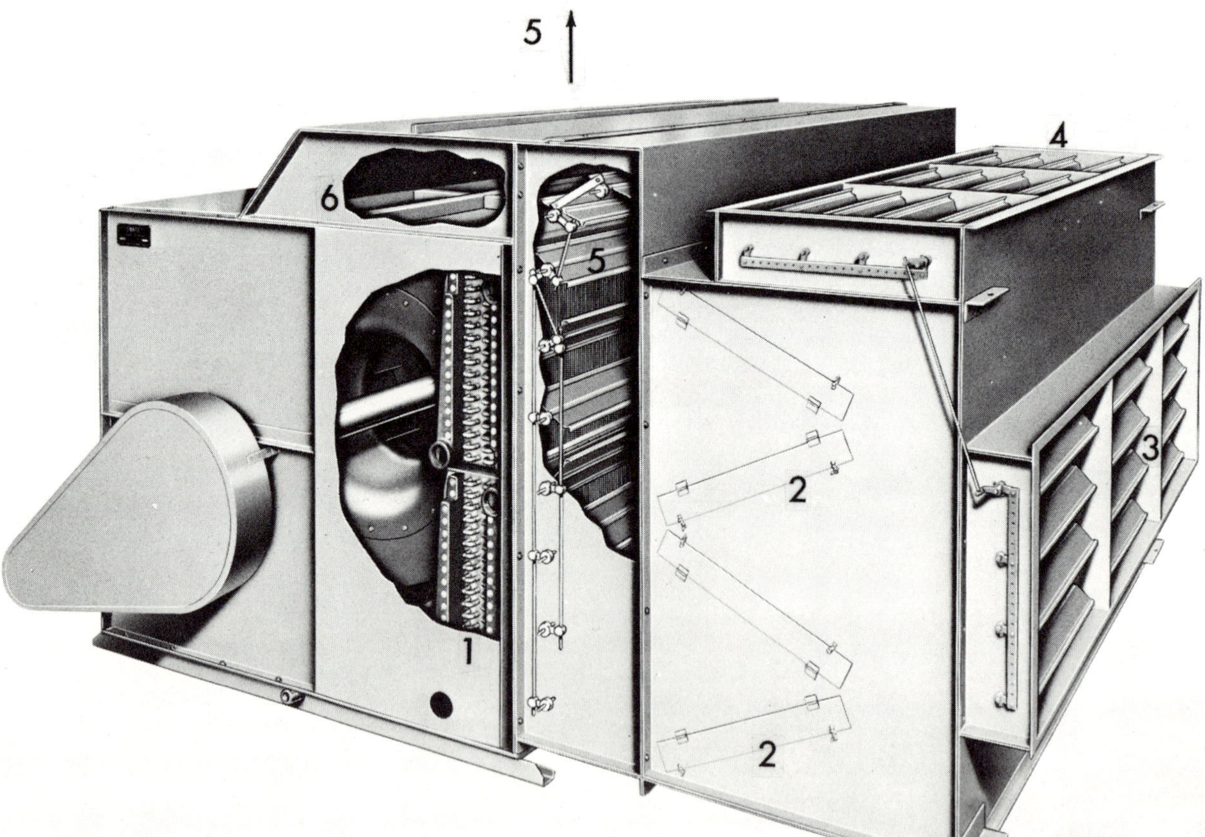

Courtesy of The Trane Company

Fig. 6-4. Typical medium-size air handler. (1) Coils; heating-cooling, or both; (2) Sectional filters inside mixing box; (3) Fresh-air intake damper; (4) Return-air damper; (5) Face and by-pass dampers; and (6) By-pass deck.

High Velocity Induction System

Figure 6-3 shows a typical high-velocity (pressure), induction system. The units are usually placed under each window, as shown, but can also be used as horizontal units suspended from the ceiling. The system can also be utilized within duct work.

Operation — The central conditioning unit draws in outside air, conditions it, and then discharges it into high-velocity ducts serving the individual induction units. The air is then discharged at high velocity through a number of small nozzles inside the unit. The high velocity of the air from the nozzles, induces (draws) air from the room through a secondary coil, containing either cold or warm water, depending on whether the units are operating on summer or winter cycle.

The air from the central air-conditioning unit furnishes the ventilation, required dehumidification, and the motive power for inducing the second-

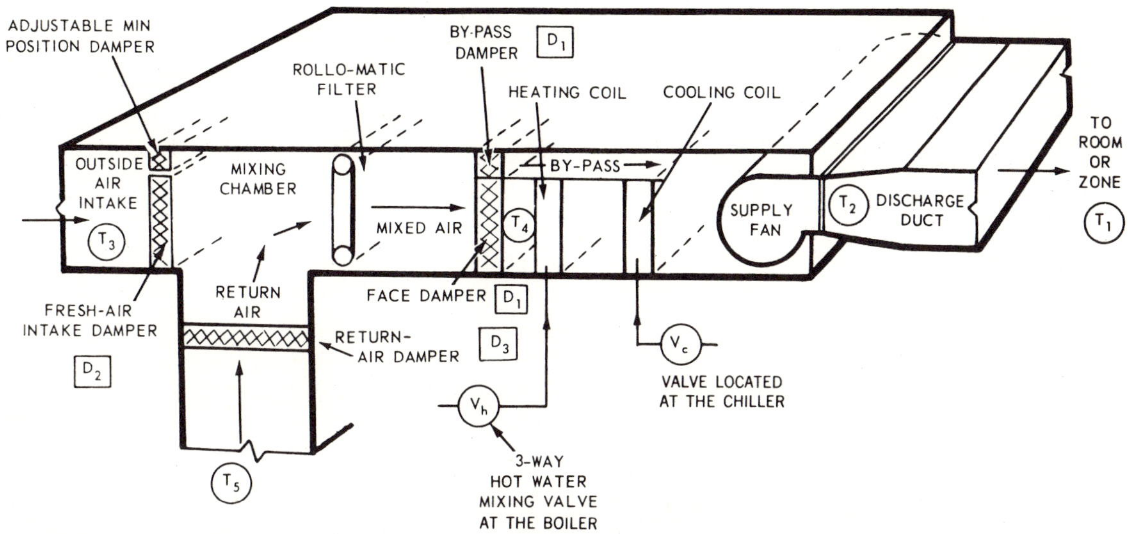

T_1 ROOM OR ZONE THERMOSTAT—Located in a room or zone. Signals back to T_2 for required action when temperature varies from set temperature in the room or zone served.

T_2 SUB-MASTER THERMOSTAT—Located in discharge duct. Upon signal from room or zone thermostat (T_1), the T_2 submaster may activate dampers D_1, D_2, and D_3 in an effort to bring temperature at T_1 back to set conditions.

T_3 HIGH LIMIT THERMOSTAT—Located in fresh-air intake. Acts to close D_2 during cooling cycle when temperature exceeds set point, normally 72°F.

T_4 FREEZSTAT—Located ahead of coils. Acts to limit travel of D_2 to prevent inrush of air, normally below 35°F.

T_5 FIRESTAT—Located in return duct. Acts to shut down unit when temperature reaches set point, normally 135°F.

D_1 FACE AND BY-PASS DAMPERS—The By-pass Damper serves the top deck; face damper serves heating and cooling coil deck below. By-pass damper is normally closed (N.C.); face damper is normally open (N.O.). When one modulates to open position, the other modulates to closed position.

D_2 FRESH-AIR INTAKE DAMPER—Controls quantity of fresh air drawn from outside. D_2 is normally closed (N.C.) when a minimum-position damper (D_{2A}) is provided. If no D_{2A} provided, D_2 also serves as a minimum-position damper.

D_{2A} MINIMUM-POSITION DAMPER—When provided, it automatically opens to min. set position when fan is cut on and D_2 remains closed until activated by T_2.

D_3 RETURN-AIR DAMPER—With fan running, it automatically assumes its set open position. With fan not running, it is normally closed.

V_h THREE-WAY HOT WATER MIXING VALVE—Modulates from completely closed to the boiler at outside temperature above 65°F to fully open to the boiler at 0°F. For temperatures in-between, it mixes boiler and return water.

V_c CHILLED-WATER VALVE—Is open only when the chilled-water pump operates which would be when operating on Mechanical Cooling Cycle.

WATER FLOW—Normally no water flows in *cooling coil* when operating on Heating Cycle, and no water flows in *heating coil* when operating on Mechanical Cooling Cycle.

Fig. 6-5. Typical single-zone, large-capacity air handler.

ary air. The induced air performs only sensible cooling in the summer and furnishes the necessary heating in the winter.

Advantages and Disadvantages — This system is becoming popular but sufficient information is not available at this time to list any number of advantages or disadvantages. It does consistently provide for cooling and/or heating the majority of the time. Since the changeover point (temperature) is not the same for all sides of a building, the system requires zoning. Because of the high air-velocity, care must be exercised to prevent entrainment of water in the supply air, particularly on days of high humidity.

Central Air Handler Serving Ceiling Diffusers

Figure 6-4 shows a typical, medium-size factory-assembled air handler. Figures 6-5 and 6-6 show

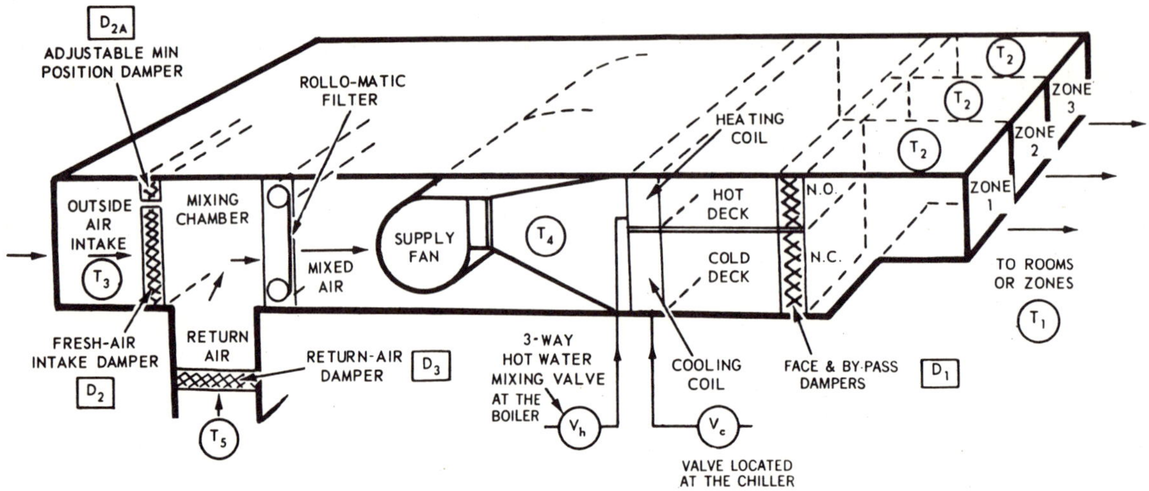

T_1 ROOM OR ZONE THERMOSTAT—Located in room(s) or zone(s). Signals back to T_2 for required action when temperature varies from set temperature in room(s) or zone(s) controlled.

T_2 SUB-MASTER THERMOSTAT—Located in discharge duct of each zone. Upon signal from its respective T_1, T_2 will act to modulate *close* either of the By-pass dampers (depending on whether warmer or cooler air is desired) while *opening* other damper in order to satisfy its respective T_1.

T_3 HIGH LIMIT THERMOSTAT—Located in fresh-air intake. Acts to close D_2 during cooling cycle when temperature exceeds set point, normally 72°F.

T_4 FREEZSTAT—Located ahead of coils. Acts to limit travel of D_2 to prevent inrush of air, normally below 35°F.

T_5 FIRESTAT—Located in return duct. Acts to shut down unit when temperature reaches set point, normally 135°F.

D_1 FACE AND BY-PASS DAMPERS—One serves the Hot Deck and other serves the Cold Deck. The hot-deck damper is normally open (N.O.); the cold-deck damper is normally closed (N.C.). When one modulates to open position, the other modulates to closed position.

D_2 FRESH-AIR INTAKE DAMPER—Controls quantity of fresh air drawn from outside. D_2 is normally closed (N.C.) when a minimum-position damper (D_{2A}) is provided. If no D_{2A} provided, D_2 also serves as a minimum-position damper.

D_{2A} MINIMUM-POSITION DAMPER—When provided, it automatically opens to min. set position when fan is cut on and D_2 remains closed until activated by T_2.

D_3 RETURN-AIR DAMPER—With fan running, it automatically assumes its set open position. With fan not running, it is normally closed.

V_h THREE-WAY HOT WATER MIXING VALVE—Modulates from completely closed to the boiler at outside temperature above 65°F to fully open to the boiler at 0°F. For temperatures in-between, it mixes boiler and return water.

V_c CHILLED-WATER VALVE—Is open only when the chilled-water pump operates, which would be when operating on Mechanical Cooling Cycle.

WATER FLOW—Normally no water flows in *cooling coil* when operating on Heating Cycle; and no water flows in *heating coil* when operating on Mechanical Cooling Cycle.

Fig. 6-6. Typical multi-zone, large-capacity air handler.

the larger type of air handler which is usually fabricated at the installation. It is large enough for a man to enter and service all the components of the unit. The large unit, as shown in Figs. 6-5 and 6-6, will be covered herein, in detail.

Components — The components of a central air handler consist of: (1) a fan, (2) a heating and cooling coil, (3) fresh-air intake and damper, (4) return chamber and damper, (5) filter(s), and (6) one or more ducts through which the cooled (or heated) air is directed to the area served by the unit.

Function — A typical air handler performs the following:

a. Circulates a mixture of return air and outside air
b. Cools or heats the air in accordance with the temperature control requirements
c. Filters the air
d. Provides cooling during the cool weather seasons without mechanical refrigeration by admitting up to 100 per cent outside air through the unit.

Types — There are two principal types of large-capacity central air handlers; single-zone and multi-zone. Their operation is similar as regards control and purpose. They differ mainly as follows:

a. *Single-zone central air handler.* (See Fig. 6-5.)
   1. Serves only one zone
   2. The heating and cooling coils are in tandem
   3. Face and by-pass dampers are in front of the coils
   4. Has by-pass deck
   5. Fan is located behind the coils.
b. *Multi-zone central air handler.* (See Fig. 6-6.)
   1. Serves two or more zones
   2. Heating coil is located in a hot deck separated from the cooling coil which is located in the cold deck below.
   3. Face and by-pass dampers are behind the coils.
   4. The hot deck and cold deck are used as by-pass decks.
   5. Fan is located ahead of the coils.

Advantages:

a. When installed in existing buildings, can utilize existing perimeter radiation for supplemental heating.
b. Under normal conditions, humidity can be controlled within satisfactory limits through coil temperature.
c. Can use outside air for cooling during intermediate seasons in lieu of operating the mechanical refrigeration plant, with resultant savings.
d. Air handler can be located in basement or penthouse, in machine room or adjacent thereto; in fact, almost anywhere to suit building requirements. Thus, they can be located in non-income-producing or needed work areas.
e. In a large building, the number of air handlers required is very much less, compared to the number of fan-coil units required. This is particularly true in large office and apartment buildings. Thus, the man-hours required in travel-time alone, for inspection and preventive maintenance, are less. In some cases air handlers can be grouped in the same room.
f. Air handlers with separate heating and cooling coils are more flexible in operation and easier on the refrigeration machine at time of changeover from heating to cooling.

Disadvantages:

a. The operating controls are usually complex and not fully understood by the average maintenance employee.
b. Cold or excessive drafts from discharge grilles are sometimes hard to correct and can waste many man-hours in trying to locate and correct the condition.
c. A malfunction in outside fresh-air-intake damper, or in the controls, which allow this damper to stay open to any degree when it should be closed, can result in excessive fuel costs in the winter and high utility costs in the summer.

d. When perimeter heating equipment is utilized to implement heating with air handlers, the two are sometimes very hard to control properly.
e. Usually requires maintenance personnel having some mechanical ability.
f. Requires duct-work that makes installation in existing buildings difficult, expensive, and very disrupting to occupants.
g. Cooling and heating all-year-round is possible only through complicated systems requiring real knowledge on the part of operating personnel.
h. In large installations, proper balancing of the system requires the services of mechanics specially trained in this field.
i. Walls and ceilings around discharge grilles usually show discoloration.

Operation

We will use the single-zone air handler to explain in detail how the temperature of the air leaving the discharge duct of the unit is controlled by the automatic repositioning of the dampers activated by the sub-master thermostat, T_2, in the discharge duct, in response to signals from the room or zone thermostat, T_1.

T_1 - T_2 Relationship

For each room or zone served by an air handler, there is a specific T_1 - T_2 relationship table based on design requirements for that particular room or zone. Assume a T_1 - T_2 relationship for a typical zone as follows:

T_1	T_2
72°F	85°F
74°F	75°F
76°F	65°F
78°F	55°F

This means that to produce a temperature of 74° F in this zone, the temperature of the air as it is being discharged from the air handler (where T_2 is located) must be 75° F; to produce 78° F, the discharge temperature would have to be 55° F: This relationship was originally determined at time of design and was supposed to have been checked out at time of installation, and the final table shown on the drawing furnished for a particular air handler, should be correct.

Face and By-pass Dampers (D_2) — The repositioning of the face and by-pass dampers is accomplished through the T_2 sub-master thermostat in response to signals from the room, or zone thermostat T_1, as the temperature of the room or zone changes from set temperature.

Example of a Heating Cycle — Assume we are in the heating season (heating cycle), the set point for the zone is 74° F and the same T_1 - T_2 table applies. Now assume the temperature in this zone had risen to 75° F. The T_2 sub-master would then position the face and by-pass dampers (D_1), by a gradual closing of the face damper serving the heating coil and a gradual opening of the by-pass damper serving the by-pass deck, to produce a discharge temperature at T_2 of 70° F. If the temperature continued to rise, say to 76° F, the face damper to the heating coil would be fully closed, while the damper to the by-pass deck would be fully open, to produce a discharge temperature at T_2 of 65° F.

Fresh-Air-Intake Dampers

1. *Minimum-position Damper* (D_{2A}) — This damper, when provided, is always open to its set position when the fan is operating and is closed when the unit is not operating. It supplies the minimum amount of fresh air necessary to off-set infiltration.

   NOTE: Where no minimum position damper such as D_{2A} has been provided (usually in older installations), the fresh-air-intake damper (described next) serves this purpose.

2. *Fresh-Air-Intake Damper* (D_2)—The damper is activated by the T_2 sub-master when the action of the by-pass dampers (D_1) is not sufficient to maintain the zone temperature at design or set conditions.
   Example — Take the same heating example given above, and assume the temperature in the zone had now risen above 76° F. The T_2 sub-master would now activate the

fresh-air-intake damper (D_2) to modulate open while causing the return-air damper (D_3) to modulate closed until the set temperature in the zone was satisfied. If the zone then proceeded to cool off, the dampers would begin to act in reverse.

3. *Combined Damper Position* — When the fresh-air-intake damper, (D_2), *also* serves as the minimum-position damper, the damper is closed when the unit is not operating. When the fan is cut on, the damper automatically opens to its set minimum position and remains in that position until activated by the T_2 sub-master thermostat as described in the preceding example.

Return-air Damper (D_3) — This damper is normally closed when the unit is not operating. When the fan is cut on, the damper automatically assumes its set position. It remains in this position until activated by the T_2 sub-master thermostat in conjunction with the outside fresh-air-intake damper (D_2). As this damper (D_2) modulates open, the return-air damper (D_3) modulates closed. When D_2 is wide open, D_3 is fully closed.

Mechanical Cooling Cycle — For example, assume we are in the cooling season instead of the heating season and the plant is being operated on the mechanical cooling cycle instead of the heating cycle. This means that the refrigeration plant is being operated and chilled water is circulating through the cooling coil in the air handler. When operating thus, the face and by-pass dampers perform to hold the temperature in the zone at the set point.

Outside Limit Thermostat (T_5) — In the case of cooling, this thermostat will cause the outside fresh-air-intake damper to return to minimum opening whenever the outside temperature exceeds a set point, normally, 72° F.

Ventilation Cycle — Operation under this cycle, when neither heating nor mechanical cooling is required, is accomplished through the T_1 - T_2 relationship in the same way as the two examples given. The only difference is that there will be no hot water in the heating coil or chilled water in the cooling coil. Thus, all effective control would be through the fresh-air-intake damper (D_2), as it is activated by the sub-master (T_2), from signals received from the zone thermostat (T_1). In some localities the outside Dry-Bulb Temperature may be suitable for cooling the space but the elevated Wet-Bulb Temperature would create an undesirable condition. For ultimate use of this cycle, a wet-bulb controller or a total heat sensor should be used rather than a dry-bulb controller only.

Hot Water Temperature Modulation

Controls — For installations using a hot water coil (rather than a steam coil), the temperature of the water circulating through the heating coil is varied inversely with the outside temperature through the use of an outside control and a 3-way water-mixing valve.

Outside Control — The outside control activates a water-mixing valve whereby the amount of water taken from the boiler is mixed with the return water to produce a circulating water temperature inversely proportional to the outside temperature. This means that the higher the outside temperature, the lower the temperature of the circulating water in the heating coil; and the lower the outside temperature, the higher the circulating water temperature.

Three-way Mixing Valve (V_h) — This water temperature change is accomplished through a 3-way mixing valve. The valve is so constructed that it can: (1) take water directly from the boiler (usually 190° to 210° F) and circulate it directly to the heating coil; (2) mix it with return water in any proportion and circulate the mixture through the coil; or (3) circulate only return water through the coil. The valve modulates from completely closed-off to the boiler at 65° F outside temperature, to completely open to the boiler at 0° F, outside temperature.

Water Temperature Relation Table — In the design stage, heating-plant capacity is based on the "outside design temperature" which takes into account the lowest temperature on record for a locality. In actual operation, the temperature at which the boiler water is maintained is an additional factor. For the operator of a heating plant to determine that his heating plant and related controls are operating properly, temperature relation tables are furnished. There are three tables which show the relationship between outside temperature and the corresponding temperature at which the water should be leaving the mixing valve.

Table 6-1 is based on a boiler-water-temperature maintained at 190° F; Table 6-2 for a temperature of 200°; and Table 6-3 for a temperature of 210°. Each table is set up to take care of the three most commonly used design temperatures; namely, −10°, 0°, and +10° F. All that an operator needs to do to see if his heating plant and controls are operating properly is to ascertain from the contract specifications the "design temperature" for his installation, select the proper boiler-water-temperature table and then check the temperature of the water leaving the mixing valve.

> NOTE: Where perimeter, direct type radiation is used to implement air handling units, care should be taken in providing a water temperature low enough to force the air handling units(s) to supply the final room temperature. This will avoid overheating the space by the perimeter radiation, thereby forcing the air handler units(s) to provide outside air to cool off the space.

Air Handler Equipment Checks

During Season:

1. Check bearings of motors and fan, by feel. Lubricate motor and fan bearings as recommended by the manufacturer.
2. Check fan shaft for end play, vibration, and bearing wear.
3. Check belts for proper tension, and pulleys for proper adjustment.
4. Check filters. Replace or clean when necessary, depending on type. Look for "necking," or tears.
5. Check coils. If not clean, they must be air cleaned, flush cleaned, or steam cleaned, depending on condition.
6. Check operation of dampers, automatic valves, and linkages. Make certain minimum position, fresh-air-intake damper is not open beyond its pre-set design position (usually 10 to 15 per cent open) unless called for. Make certain main, outside fresh-air-intake damper is not open when face damper of heating or cooling coil is not *fully* closed.
7. Check outside air-intake screens; clean if necessary.

End of Season:

1. Clean and flush condensation pan and its drain lines.
2. Check fan blades for looseness and dirty condition. Clean if necessary.
3. Check all junction boxes and contactor connections on fan motors for tightness, by qualified personnel. Check condition of contactor points. Have replacements made as needed.
4. Where possible, observe entire duct system for loose or missing insulation, openings, caved-in duct, fire dampers, etc., especially on new installations.
5. Clean all damper leakages with a solvent. Lubricate with graphite and penetrating oil.
6. List any needed repairs or replacement of worn out parts.
7. Note any signs of rusting or the need of touch-up painting.

Equipment Checks
(Pneumatic Compressor for Controls)

During Season:

1. Check air pressure in supply tank and the pressure regulator adjustment in supply line, for proper limits.
2. Check air-compressor oil level; if low, add oil.
3. Check belt tension and alignment.
4. Drain water from air tanks and supply-line filters.
5. Test air-relief valve per manufacturer's instructions.
6. Clean air-compressor intake filter pads.
7. Check all mounting screws and bolts for proper tightness.

Table 6-1. Water Temperature Relation Table, °F

Outside temperature °F	Temperature of water leaving mixing valve		
	Design temp. -10°F	Design temp. 0°F	Design temp. +10°F
Above 65°F	Circulating pump off. No water flowing		
65	80	80	80
60	88	89	90
55	95	97	100
50	102	106	110
45	110	114	120
40	117	122	130
35	124	131	140
30	131	139	150
25	139	148	160
20	146	156	170
15	153	164	180
10	161	173	190
5	168	181	-
0	175	190	-
-5	183	-	-
-10	190	-	-

190 Deg. F Boiler Temperature Water

Note: For buildings having a large, internal heat gain due to high-intensity lighting or a large number of motors, it may be desirable to maintain a much lower boiling-water temperature during the early and late months of the heating season to avoid overheating.

Table 6-2. Water Temperature Relation Table, °F

Outside temperature °F	Temperature of water leaving mixing valve		
	Design temp. -10° F	Design temp. 0° F	Design temp. +10° F
Above 65° F	Circulating pump off. No water flowing		
65	80	80	80
60	88	89	90
55	96	99	102
50	104	108	113
45	114	118	125
40	120	126	134
35	128	135	145
30	136	145	156
25	144	154	167
20	152	163	178
15	160	172	189
10	168	181	200
5	176	190	-
0	184	200	-
-5	192	-	-
-10	200	-	-

200 Deg. F Boiler Temperature Water

Note: For buildings having a large, internal heat gain due to high-intensity lighting or a large number of motors, it may be desirable to maintain a much lower boiling-water temperature during the early and late months of the heating season to avoid overheating.

Table 6-3. Water Temperature Relation Table, °F

Outside temperature °F	Temperature of water leaving mixing valve		
	Design temp. -10° F	Design temp. 0°F	Design temp. +10° F
Above 65°F	Circulating pump off. No water flowing		
65	80	80	80
60	89	90	92
55	98	100	103
50	106	110	115
45	115	120	127
40	123	130	139
35	132	140	150
30	141	150	162
25	150	160	174
20	158	170	186
15	167	180	198
10	176	190	210
5	184	200	-
0	193	210	-
-5	201	-	-
-10	210	-	-

210 Deg. F Boiler Temperature Water

Note: For buildings having a large, internal heat gain due to high-intensity lighting or a large number of motors, it may be desirable to maintain a much lower boiling-water temperature during the early and late months of the heating season to avoid overheating.

8. Check all connections for air leaks using a soap solution. If air-compressor runs excessively, check for loss of air pressure at controls and throughout piping system.

End of Season:

1. Change air-compressor crankcase oil.
2. Oil electric-motor bearings per manufacturer's recommendations.
3. List any needed repairs or replacements of worn out parts.

Air-Handler Log (Record) — It is desirable that an operating record (log) of the air handlers be maintained. The log will provide a record of the condition of the filters and thus indicate when replacements are needed; it also will indicate any improper operation of the air handlers. (See Exhibit 6-A.)

Air-Handler Diagram — Where more than one air handler is installed, it is a decided advantage to have a line diagram prepared for each air handler. The diagram should be placed under glass and posted on the wall near its respective unit. The diagram should show the number of the unit for identification, pertinent data, such as: CFM capacity, heating and cooling-coil data, motor rating, etc., as well as the specific building areas served by the unit. The diagram should also show, in concise and easily understood terms, all necessary operating and maintenance instructions for the unit. In the case of new construction, the preparation of these diagrams and instructions should be made a part of the original specifications. For other installations, it is suggested that arrangements be made with a qualified engineer to prepare them. The cost should be nominal compared to the advantages gained from improved operation, maintenance, trouble shooting, down-time, and the quicker training of new maintenance personnel.

Winter Lay-up of Cooling Coil — During winter shut-down, those cooling coils subject to freezing must be blown out. Even though these coils are drained, some water (which could freeze) may remain in the coils with resultant damage to the coil. After making certain all water is out of the coils, also be sure all plugs are out and drains left open. Leave coils thus until start-up next spring.

Another method is to drain the coils and then flush them with an anti-freeze. In the process, the anti-freeze may be reclaimed for use in flushing additional coils as long as the proper strength of anti-freeze is maintained. It is not necessary to lay-up the coils full of anti-freeze since any liquid remaining in the coils will be mixed with anti-freeze.

> NOTE: Where no pre-heat coil for outside air has been provided, the cooling coil should always be placed on the downstream side of the heating coil. This will prevent freezing of the cooling coil from minimum outside air temperature in cold weather.

Air Balance

Where central air-distribution systems are utilized, it is advisable to have the system balanced by a qualified air-balancing concern, preferably by one that is a member of the Air Balancing Council. In fact, this should be included in the specifications before acceptance of a new building.

The cost of this service at the time of construction would be small compared to the later cost when the building is already occupied and it is discovered that the system is not balanced properly and complaints are received from the occupants. This is a specialized field and requires special equipment and a thorough knowledge of controls. An unbalanced system, whether in design or from improper installation, can result in many wasted man-hours by maintenance personnel in trying to correct complaints. Further, when unqualified maintenance personnel try to adjust the controls and dampers, the system usually can become hopelessly unbalanced which will lead to further complaints and increased fuel and utility costs.

Training — The proper training of maintenance personnel in the operation and care of air handlers by factory representatives is of prime importance. Here again, the building specifications should spell out the extent of the training to be provided, the particular time it is to be given, and the number of maintenance employees to whom the training is to be given. It should also specify that the instructors are to be company employees

OPERATING LOG FOR AIR HANDLERS

BUILDING	MONTH AND YEAR

DATE	UNIT NO.	TIME	FAN		TEMPERATURES									STEAM PRESS. OR HOT WATER TEMP.	PRESS. DROP ACROSS FILTERS	CONDI-TION OF FILTERS
					AIR							CHILLED WATER				
					OUT-SIDE	RETURN	LEAVING COIL.	ZONE								
			ON	OFF				1	2	3	4	IN	OUT			

Exhibit 6-A. Operating log for air handlers.

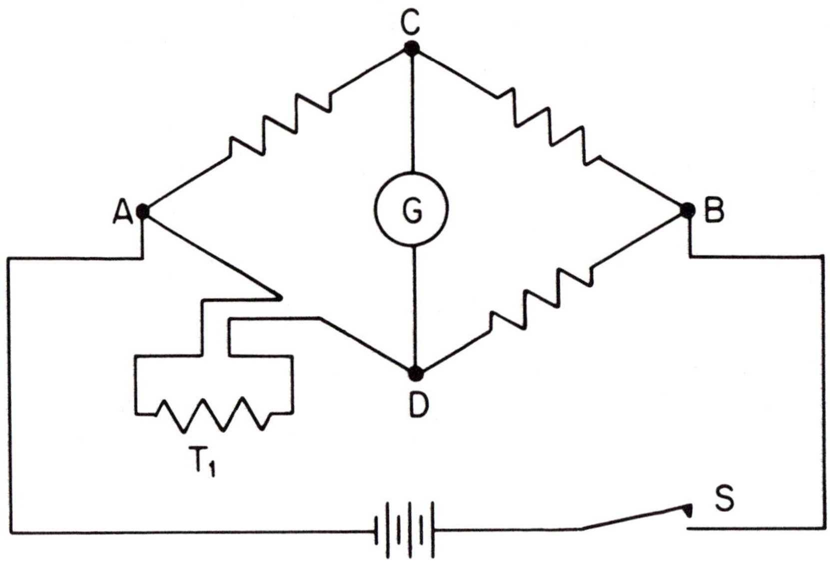

Fig. 6-7. The basic Wheatstone bridge.

trained in this kind of work and for this particular piece of equipment. Most important of all, the training should utilize the manufacturer's operation and maintenance manual for the particular equipment *and* each maintenance employee taking the instructions should be given a copy for his *own use and reference.*

Air Handler Controls

Basics — The art of controls is continually changing and can be quite complicated. Because of this, we will present only a general coverage involving a basic application.

Electronic Controls — Since electronic controls are used to a great extent in conjunction with pneumatically-operated valves, or damper motors, the operation of only this one type of control will be described here in detail. The main difference between a straight electric system and an electronic system is in the sensing element of the thermostat. In the electric system, the thermostat has either a bimetal tongue or a temperature-sensitive bulb to initiate the required mechanical action. In the electronic system, the thermostat uses a coil of wire (referred to as a resistor) instead of the bimetal tongue or bulb. The sensing element or resistor in the electronic thermostat reacts to ambient temperature changes as follows:

a. Resistance increases with an increase in temperature.
b. Resistance decreases with a decrease in temperature.

Wheatstone Bridge — Current for electronic controls is based upon the principles of the "Wheatstone Bridge." As a simple approach to understanding the bridge, one using direct current will be used to illustrate the basic principles of the bridge. (See Fig. 6-7.)

T_1 represents a room or zone thermostat. As stated, the sensing element of the thermostat is a coil of wire which acts as a resistance in the bridge circuit. Assume all the resistances in the bridge had the same value. Upon closing the switch (S), there would be no current flow through the galvanometer (G).

Temperatures

Lower Temperature — If the resistor in T_1, seen in Fig. 6-7, experiences a *lower* temperature, AD will have a *lower* resistance than AC. This will make point (D) positive with respect to point (C) and current will flow from (C) to (D) through the galvanometer (G).

Higher Temperature — If the resistor in T_1, seen in Fig. 6-7, experiences a *higher* temperature, AD will have a *higher* resistance than AC. This will make point (C) positive in respect to (D) and current will flow in the opposite direction, that is, from (D) to (C). The bridge using alternating current is basically the same except that use is made of the phase difference in alternating current. In an AC circuit, half of the current flows in one direction during one half of the cycle and, during the second half of the cycle, the other half of the current flows in the opposite direction. Through the use of amplifiers, signal strength is increased as much as 2,500 times.

T_1 - T_2 *Relationship* (See *Operation*) — Two bridges are used to accomplish the relationship between T_1 (room or zone thermostat) and T_2 (duct discharge or sub-master thermostat). One bridge is referred to as the main bridge and the other as the auxiliary bridge. (See Fig. 6-8.)

T_1 *(Room or Zone Thermostat)* — The resistance wire in the T_1 thermostat forms one leg of the main bridge. Any change in the room or zone temperature causes an unbalance in the main bridge circuit. This in turn, produces a signal of a few millivolts which is impressed across the terminals of the amplifier.

T_2 *(Discharge Duct or Sub-master Thermostat)* — The resistance wire in this thermostat is in an averaging tube that senses temperature variation across the discharge duct. The sensing element forms one leg of the auxiliary bridge. This bridge is connected in parallel with the main bridge and the two, in turn, are connected by a ratio adjustment potentiometer.

Ratio Balancing Potentiometer — The potentiometer in the circuit determines the number of degrees change T_2 must make for each degree change in T_1. It is connected with the amplifier so it will balance-out the incoming signal, both as to phase and strength. It is this potentiometer that makes the damper motor capable of proportional operation. The adjustment is determined by design and operating conditions and, upon installation, further adjustment is made by the control company, if necessary.

Balanced Circuit — When the temperature of the discharge air is such that space conditions are satisfied, the voltage-drops across the two bridges arrive at a balance. With no signal across the amplifier terminals, "nil point" is reached.

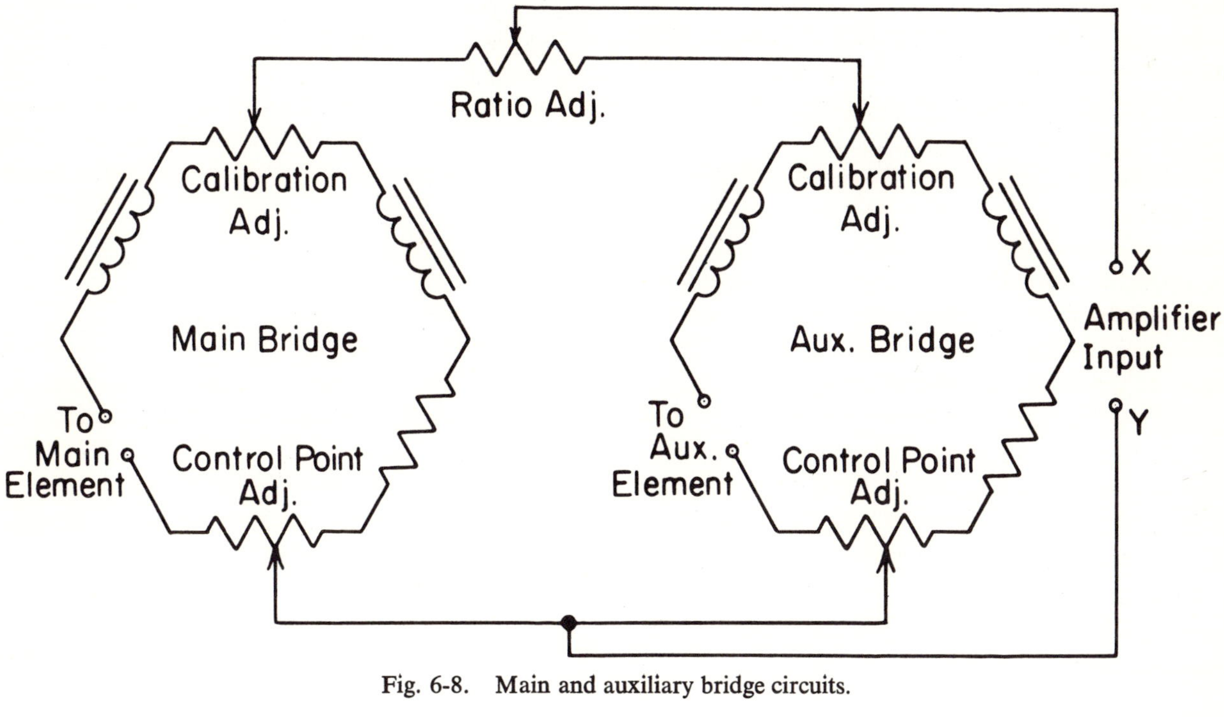

Fig. 6-8. Main and auxiliary bridge circuits.

Single-Zone Cycle Control. (See Fig. 6-5.)

Heating-Cycle

Water Pumps — The chilled-water pump is off. No water is flowing through the chilled-water coil. The hot-water pump is running heated water from the boiler through the hot-water coil.

Call for Heat — As the temperature in the room (represented by T_1) decreases, there is a *decrease* in resistance in T_1. Thus, a *negative* phase signal is impressed across the amplifier. T_2 will be calling for sufficient *heat* to satisfy space requirements and rebalance the bridge system. The dampers will react as follows:

1. D_1 (*Face and By-pass Dampers*) — Decreasing voltage on the solenoid coil will start opening the restrictor on the oil by-pass which, in turn, will *lower* the pressure on the piston and start the face damper modulating toward the *open* position.
2. D_3 *(Return Damper)* — This damper will modulate toward the open position. This is a shaded pole motor that is reversible.

Overheating — As the temperature in the room (represented by T_1) increases, there is an *increase* in resistance in T_1. Thus a *positive* phase signal is impressed across the amplifier. T_2 will be calling for sufficient *cooling* to satisfy space requirements and rebalance the bridge system. The dampers will react as follows:

1. D_1 (*Face Damper*) — Increasing voltage on the solenoid coil in D_1 will start closing the restrictor on the oil by-pass which, in turn, will *raise* the pressure on the piston and start the face damper modulating toward the *closed* position. As continued cooling is needed, D_1 will modulate the face damper to a completely closed position with the bypass damper fully open.
2. D_2 (*Outside Fresh Air Damper*) — Should more cooling be needed to bring the room temperature (T_1) down, the outside fresh-air damper motor will start the damper to modulate toward the open position as soon as the face damper is *fully closed*. The motor is a shaded pole, reversible type, with the shaft spring loaded to close the damper in case of power failure.
3. D_3 (*Return Damper*) — In an overheated situation, the return damper will modulate toward its minimum open position. The return damper will remain in this minimum position as long as the fresh-air-intake damper is open.

Mechanical Cooling Cycle

Water Pumps — The hot-water pump is normally off. No water is flowing through the hot water coil. The chilled-water pump is running chilled water from the refrigeration plant through the chilled-water coil.

Call for Cooling — As the temperature in the room (represented by T_1) increases, there is an *increase* in resistance in T_1.

Thus a *positive* phase signal is impressed across the amplifier. T_2 will be calling for sufficient *cooling* to satisfy space requirements and rebalance the bridge system. The dampers will react as follows:

1. D_1 (*Face and By-pass Dampers*) — Increasing voltage on the solenoid will start opening the restrictor on the coil by-pass which, in turn, will lower the pressure on the piston and start the face damper modulating toward the open position and the by-pass to the closed position. In the single-zone unit, the reversing relay reverses connections to the motor's starting windings so it has the same sequence of operation on the face and by-pass dampers in both heating and cooling seasons. This is only true for the single-zone unit.
2. D_2 (*Outside Fresh-air Damper*) — When operating under the mechanical cooling cycle, this damper is under control of the outside limit thermostat T_3. Since the setting of T_3 is normally 72° F, damper D_2 will be in minimum position (or closed position when separate minimum-position damper is provided) when outside temperature exceeds 72° F.
3. D_3 (*Return Damper*) — Since D_3 operates

inversely in relation to D_2, damper D_3 will be in its normal open position.

Overcooling — As the temperature in the room (represented by T_1) drops, there is a *decrease* in resistance in T_1. Thus, a *negative* phase signal is impressed across the amplifier. T_2 will be calling for sufficient heat to satisfy space requirements and rebalance the bridge·system. The dampers will react as follows:

1. D_1 (*Face and By-pass Dampers*) — Decreasing voltage on the solenoid coil D_1 will start closing the restrictor on the oil by-pass which, in turn, will *raise* the pressure on the piston and start the face damper modulating toward the *closed* position. As continued heating is needed, D_1 will modulate the face damper to a completely closed position with the by-pass damper fully open.
2. D_2 (*Outside Fresh Air Damper*) — Remains in minimum or closed position.
3. D_3 (*Return Damper*) — Remains in open position.

Outside Cooling Cycle

When Used — This cycle is used when the building can be cooled with outside air instead of using the mechanical cooling plant. This is particularly true during the late fall and early spring seasons, and certain days and nights during the summer. Under this cycle, there will be no hot water flowing through the heating coil or chilled water in the cooling coil. In the interest of economy, this cycle should be used whenever possible.

Damper Control—This is accomplished through the T_1 - T_2 relationship. With the refrigeration plant shut off, the fresh-air damper D_2 is under control of the T_2 sub-master thermostat in the discharge duct. Since the outside Dry-Bulb Temperature will be less than 72° F, the outside fresh-air-intake damper may modulate open upon a call for cooling from T_1, the room thermostat.

Damper Action — The action of the face and by-pass dampers D_1, the fresh-air-intake damper D_2 and the return damper D_3 will be the same as that described for Overheating when operating under the Heating Cycle.

Multi-Zone Cycle Control. (See Fig. 6-6.)

The action of the controls for the multi-zone unit is the same as that for the single-zone with the exception of the face and by-pass damper D_1. In the multi-zone unit, there is no reversing relay.

Damper D_1 performs the same for both the heating and mechanical cooling cycles.

CHAPTER 7

Filters

Sectional Throwaway (Glass Fibre) Filters

Where this type of disposable filter is used, they should be removed, discarded, and replaced with new ones when dirty. They may be vacuum cleaned and re-used temporarily, *only* in an emergency and until a replacement is available.

Advantages:

1. Low initial and replacement cost compared to Cleanable and Roll-type
2. One unskilled man (a janitor or his helper) can install and replace these filters
3. Readily obtainable
4. Can be used in both small and large installations.

Disadvantages:

1. Requires frequent manual check for condition of filter media
2. More frequent replacement needed as compared to Cleanable and Roll-type.
3. Does not filter as evenly over a period of time as the Roll-type.
4. Where large numbers are installed, man-hours required to service the filters may outweigh the low installation and replacement costs.

Sectional Cleanable (Coated) Filters

Where this type of coated filter is used, they should be washed, periodically, in a solution of mild detergent and water. A supply of spare filters should be kept on hand so that the air handlers need not be shut down while the filters are being cleaned. For installations with a large number of filters, a contract for cleaning and servicing them should be considered.

Advantages:

1. Initial cost is less than with the Roll-type.
2. One unskilled man (janitor or helper) can install, service, and replace this type of filter.
3. Initial installation will normally last for years.
4. Easier to obtain than the Roll-type.
5. Can be used in both small and large installations.

Disadvantages:

1. Requires frequent manual check for condition of filter.
2. Frequency of *cleaning* normally same as frequency of *replacement* for the throwaway, glass fibre type.
3. Cleaning and re-coating is a messy and time-consuming operation.
4. Where large numbers are installed, the many man-hours required to service them may outweigh the higher initial and replacement cost of the Roll-type filter.

Roll-type (Glass Fibre) Filters

This type of filter is coated with an adhesive and packaged in compact rolls 65 to 70 feet in length and from 2 to 6 feet in width. The rolls are mounted in a specially designed frame within an air handler. (See Fig. 7-1.) The frame provides for automatic unwinding of the roll in accordance with the amount of dust and dirt generated in the space served by the air handler and deposited on the filter media. The advancement of the media is accomplished by one of the following:

1. *Timer* — Advances the media in increments at a predetermined rate.

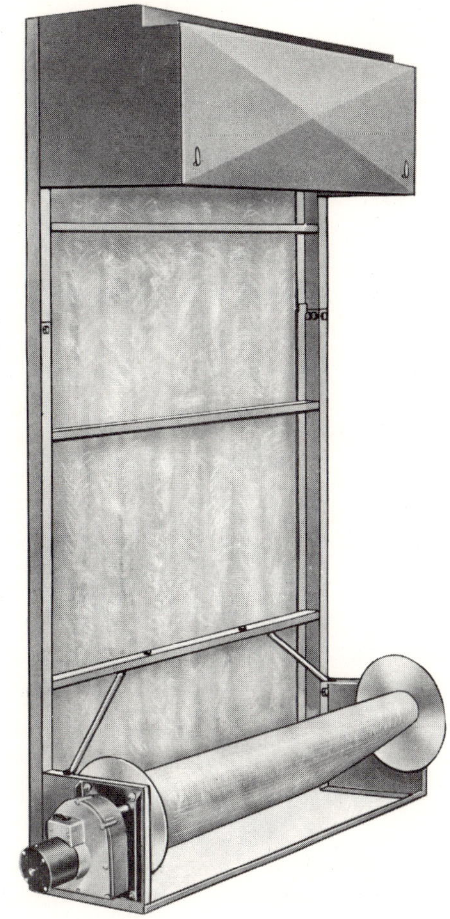

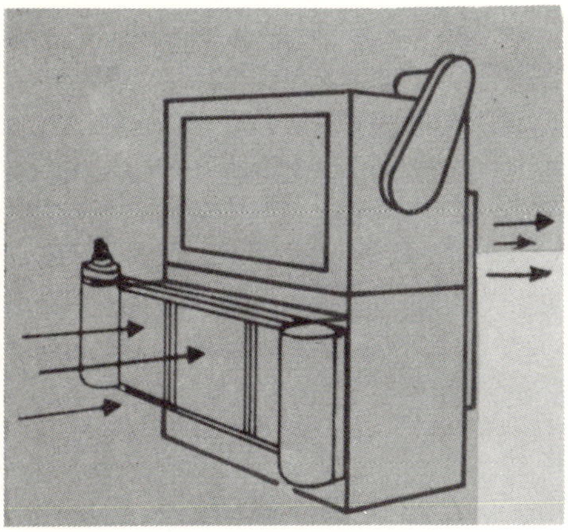

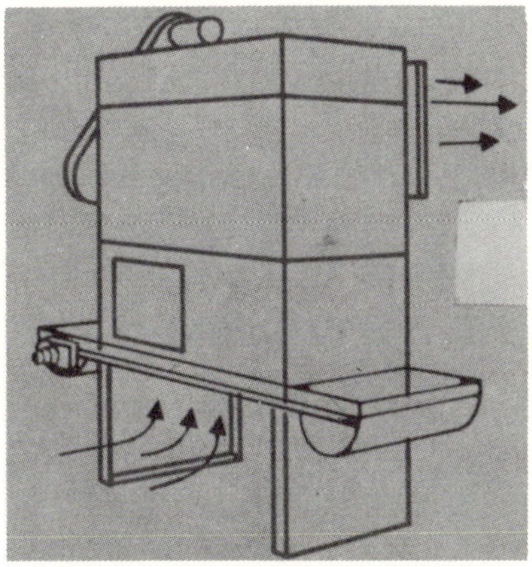

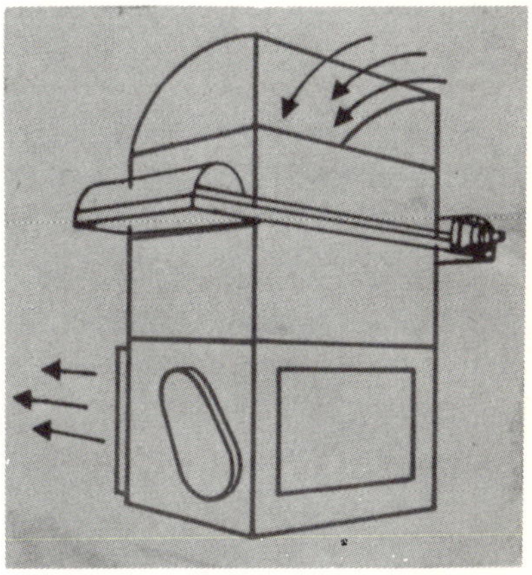

Courtesy of the Farr Company

Fig. 7-1. Roll type filters. (Top) Air flows through progressively loading media in an automatic, vertical, renewable media air filter. (Below) Horizontal units are also available, to fit various installations.

2. *Pressure Switch* — Advances the media when the build-up resistance across the media exceeds a set pressure. The pressure range is usually between .25 and .45 inch of water as shown on the gauge that comes with the control.
3. *Combination Timer and Pressure Switch* — Operates as described in (1) except that a pressure switch, as described in (2) will override the timer if an unusually dusty condition causes the pressure drop to rise to an excessive value before the timer calls for an advancement of the media.

Advantages:

1. Automatic operation as the media advances in accordance with the varying dust and dirt conditions in the space served.
2. Least man-hours required for checking condition of filter and media.
3. Longer period between replacements than sectional type due to length of rolls.
4. Does a more even job of filtering than the Sectionals over a period of time.
5. Favorable in large installations having heavy contamination loads where the replacement cost of Sectional Filters would be prohibitive.

Disadvantages:

1. High initial and replacement cost. ($30 to $50 per roll depending on length and width of roll.)
2. Filter installation and removal much more difficult and requires more time than Sectionals. From 1 to 2 men will require 30 to 50 minutes to change a roll, depending on its size. Some of the larger rolls may weigh in excess of 45 lbs.
3. Requires some know-how on the part of maintenance personnel when replacing rolls. Also, automatic advancing mechanism requires occasional attention. When this happens, the roll may run too slowly, too fast, or stop altogether. It is not an easy matter to match the rate of advance to the conditions in the space served. Also, in some cases when a pressure switch alone is used, the amount of dust or dirt generated may be so little as to result in the drying out of the adhesive on the media before the pressure differential becomes sufficient to activate the mechanism.

Special Sectional Filters

The sectional filters referred to previously, are more or less the familiar, conventional type. Figure 7-2 shows several "special" sectional-type filters. The "cube-type" filter offers five filtering surfaces compared to the usual single surface. Figure 7-2 also shows a filter having the depth of a cube type, but with a measurable, effective, filtering surface of more than 4 times that of the cube type. The filter media for both types is made from a special modacrylic fibre and is furnished with either a dry or viscose coated surface. The manufacturer claims the filters are capable of high efficiency and longer service life.

Clean Filters — Clean, efficient filters are not only necessary to remove dust, lint, and other foreign material from the air, but also insure efficient transfer of heat and cold from the heating and cooling coils to the air, as it is blown across them. Where air is found to be by-passing a filter due to: poor installation, necking of the filter, or holes in the filter, the condition should be corrected immediately. Failing to do so can result in a clogged or frozen coil with resultant expense in man-hours needed to clean the coil or even to replacing it. Access to the coil is sometimes difficult, the man-hours required can be costly and in addition, there is the annoyance of having to shut down the unit for a long repair period.

> CAUTION: Under no circumstances shall air handling units be operated without any filters.

Installation Precautions — When installing filters, whether Sectional or Roll-type, there are certain precautions that must be taken. Some of these are listed below:

1. *Air Leakage* — Check to see that there is no air leaking around the filter frame(s) and that there are no tear holes in the filter

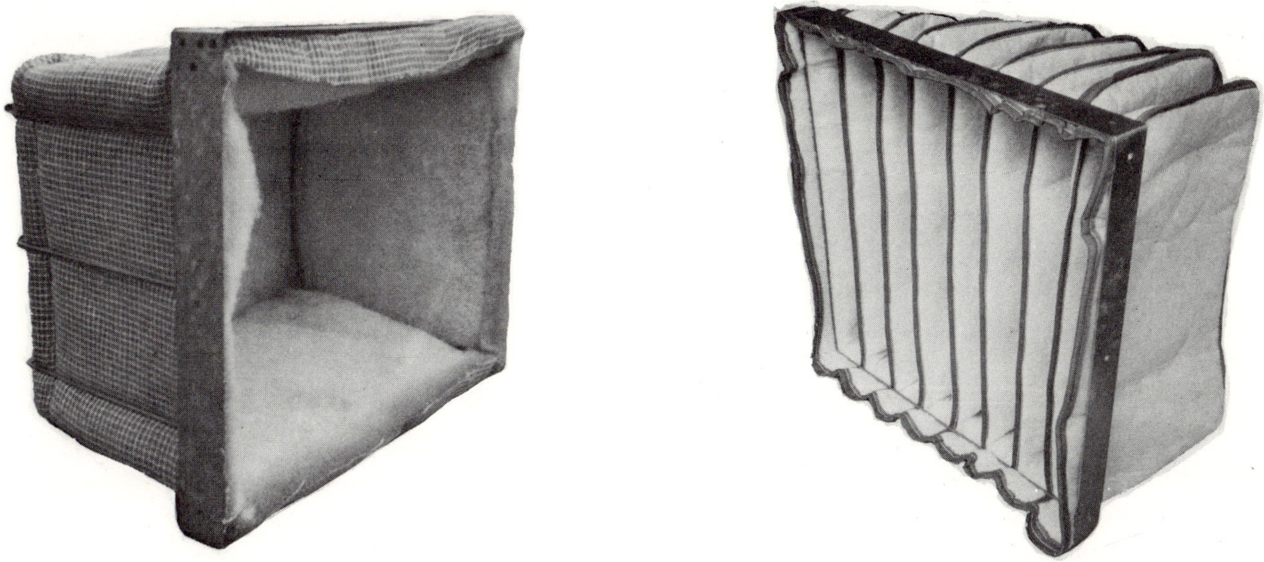

Courtesy of Union Carbide Corporation

Fig. 7-2. (Left) Panel type filter; (Right) Multi-wedge type filter.

media. All air *must* go through the filter media.

2. *Air Flow Direction* — Make certain the direction of air-flow through the filter (usually shown by arrows on the side of the filter) corresponds to the direction in which the air flows.
3. *Necking* — In the case of the Roll-type filters, be sure that the draw on the media has not stretched it to the point where it has narrowed (or necked), thus permitting air to by-pass the filter media.
4. *Advancing Mechanism* — Be sure that the mechanism is operating properly before leaving it on its own. This is particularly true in the case of automatic timers. If this is not done, it is possible that the media may run completely through in a very short time.

Filter Life — There is no way of knowing what the life expectancy of a Roll-type filter will be except under actual use. Its life will depend on the dust and dirt conditions within the space served by the air handler in which the filter is installed. Also, these conditions are not always the same for all 12 months of the year. Those serving office space only, may last from 6 to 12 months. Those serving work areas, may last as little as 10 to 20 weeks.

Specifications for Roll-type Filter — When soliciting bids for this type of filter, to be used as replacement of the original filters, it is important that the type and size are suitable. To assure this, a specification comparable to the one set forth below, for a glass fibre media should be incorporated in the bids:

> The filter roll shall contain not less than 65 feet of filter media, shall be of a glass fibre, noninflammable material, and be wound on a roll suitable for installation and operation in the filter frame of the present installation without modification. The media shall have a "scrim backing" bonded to the media, of sufficient strength to prevent necking and stretching of the media as it moves across the face of the filter. The "scrim backing" may be omitted provided the media has a built-in, separate, reinforcing material of sufficient strength to accomplish the same purpose as the "scrim backing.
>
> The media, when operating at an air-flow rate of not less than 400 f.p.m. face velocity, and when dust consisting of 96 per cent (by weight) Cottrell precipitate and 4 per cent cotton lint, is dispersed in the entering air at the rate of approximately 1 gram per 1,000 cubic feet of air, shall have received and re-

tained not less than 200 grams of dust per square foot of media face area when the average resistance to air-flow reaches 0.45 inch water gauge.

The filter shall have an average efficiency in removing Cottrell precipitate of not less than 75 per cent during the foregoing loading test.

CHAPTER 8

Preventive Maintenance

Maintenance Defined

Preventive maintenance consists of regularly scheduled inspections and lubrications along with adjustments, minor parts replacement, and making minor repairs, necessary to keep plant and equipment in good operating condition; to prevent breakdown; harmful depreciation; and costly major repairs.

Scope — Preventive maintenance may range from periodic inspections to a broad interpretation covering any activity for preventing breakdowns, cutting operating costs, and improving service. Even when well-planned, preventive maintenance is not in itself a planned program, but only a part of a completely planned or scheduled maintenance program. It is not a cure-all, but must be tied-in with all other maintenance functions.

Benefits — The full benefits of a realistic and comprehensive maintenance program cannot be realized at once. It may require several years to get it thoroughly in gear and to the point where maximum benefits can be realized. This period varies with the size and the amount of mechanical equipment involved.

Some beneficial returns from an effective preventive maintenance program are:

1. Less "down time"
2. Less overtime pay for emergency repairs
3. Less cost for minor repairs made before breakdowns
4. More dependable service
5. Extended useful life-expectancy in plant and equipment
6. Less-costly major maintenance
7. A better means of identifying causes of maintenance breakdowns and operator abuse
8. Better spare-parts control, and a smaller inventory
9. Improved safety protection — reducing "lost days" and compensation payments.

Check Lists — Check lists serve as guides in the development of a successful preventive maintenance program. Check lists are intended to cover routine preventive maintenance and only those repairs which can be classed as minor and which can be accomplished easily by the maintenance personnel. Major repairs do not come under this program.

Sample Check List

Table 8-1 has been set up as a sample check list. Only one sample check list has been set up in lieu of providing a separate check list for reciprocating, centrifugal, and absorption type machines. In applying the sample check list to a specific installation, non-applicable items are to be deleted, and any applicable items are to be added. The check list thus prepared, becomes the master check list around which all preventive maintenance performances revolve.

Applicable Installations — The sample check list is applicable to central installations (as defined in Chapter 3) having the following four basic components:

1. Refrigeration machine(s)
2. Water-cooling tower and condenser-water system
3. Chilled-water system
4. Air-handler system.

Frequency and Time — The items, frequencies, and time shown in the sample check lists, are basically for an installation having reciprocating type compressors (approximately 30 to 150 tons). Al-

Table 8-1. Typical Routine Check List (page 1).

ROUTINE PREVENTIVE MAINTENANCE SERVICE CHECK LIST				NUMBER
SYSTEM AIR CONDITIONING	EQUIPMENT Refrig. Machine Room	ORIGINAL ISSUANCE DATE	REVISION(No.)	(Date)
SUB-EQUIPMENT Refrigeration Machine		APPROVED BY (Signature and title)		

Part or Component	No.	INSTRUCTIONS	Frequency (See note below) T	D	1	2	4	12	26	52	Time
		Item 1 is applicable to all types of refrigeration machines.									
	1	Record important data on Log Form. Observe oil level in bull's eye and any visual changes in equipment appearance such as oil spots on connections or floor under equipment. Also notice unusual noise of machine operation.		x	x	x	x	x	x	x	2
			3 Times Daily: A.M.-1st Man Reporting Noon - Man in Charge P.M.-Last man Leaving								
		Items 2-11 are generally applicable to all types of refrigeration machines.									
	2	Check flow of refrigerant through sight glass. Should be smooth and without bubbles. (Where sight glass has sealed caps, the caps have to be removed when making the check. When replacing the caps, make hand tight. (Do not use wrench)					x	x	x	x	2
	3*	Check system for leaks of refrigerant and oil around shaft seal, sight glasses, valve bonnets, flange and flare connections. Also the relief valve on condenser assembly.					x	x	x	x	8
	4	Check high pressure cut-out setting. Should be at least 25 pounds below relief valve setting.					x	x	x	x	6
	5	When shutting down unit at any time, observe pressure cut-out point on low pressure control.					x	x	x	x	1
	6	With unit shut down, check following: a. Belt tension or for loose drive coupling and alignment					x	x	x	x	1
		b. Condenser line strainer ahead of condenser water pump. (Where unusual conditions exist, clean as often as required)					x	x	x	x	17
	7	Lubricate motor bearings as recommended by equipment manufacturer.								R	20
	8	Check all electrical leads for tightness of connections in all junction boxes and contactors, by qualified personnel.								x	80

*CAUTION: Freon refrigerants are relatively non-toxic. However, when checking for leaks, etc., sufficient ventilation should be assured to prevent oxygen deficiencies. Also, these refrigerants generate toxic products when heated to decomposition.

NOTES: (1) T = Tour, D = Daily, 1 = 1 Week, 2 = 2 Weeks, etc.
(2) Frequency and Time are for one machine. Time is in minutes. R = As Required.

Table 8-1. Typical Routine Check List (page 2).

ROUTINE PREVENTIVE MAINTENANCE SERVICE CHECK LIST NUMBER

SYSTEM	EQUIPMENT	ORIGINAL ISSUANCE DATE	REVISION(No.)	(Date)
AIR CONDITIONING	Refrig. Machine Room			

SUB-EQUIPMENT	APPROVED BY (Signature and title)
Refrigeration Machine	

Part or Component	No.	INSTRUCTIONS	Frequency T	D	(See notes below) 1	2	4	12	26	52	Time
	9	Electric Motors: a. Blow out motors with air under pressure when available. b. Clean and check motor starting switches using qualified personnel.								x	75
	10	Check all bolts on equipment for tightness as per equipment manufacturer's recommendation.								x	30
	11	When shutting down unit for winter, shut off liquid line valve and allow machine to pump down. Valves that have been shut off should be tagged. Check entire refrigeration system for refrigeration leaks. (Open Type Machines, see (*) below)								x	20
	12	Purge Unit (Where applicable)									
		a. Check purge compressor for normal operation and inspect belt for tension and wear. Keep motor clean.				x	x	x	x	x	1
		b. While machine is running, check for leaks indicated by (1) higher than normal purge discharge gauge pressure, (2) purge relief valve discharging to atmosphere, and (3) water accumulation in the upper bull's eye of the purge separator machine. Any water accumulation should be removed.				x	x	x	x	x	3
		c. Stop purge unit and check all fittings, flanges, etc., of the high pressure side of the unit for tightness of all connections. Make leak test, using leak - detector. Check weir standpipe, baffle plate and float valve in the separation chamber. Replace if necessary.					x	x	x	x	13

*Where there are qualified personnel, open suction and discharge valves once a month during off season and run unit a few revolutions to lubricate shaft seal. Close valves again.

NOTES: (1) T = Tour, D = Daily, 1 = 1 Week, 2 = 2 Weeks, etc.
(2) Frequency and Time are for one machine. Time is in minutes.

Table 8-1. Typical Routine Check List (page 3).

ROUTINE PREVENTIVE MAINTENANCE SERVICE CHECK LIST — NUMBER:

SYSTEM	EQUIPMENT	ORIGINAL ISSUANCE DATE	REVISION(No.)	(Date)
AIR CONDITIONING	Refrig. Machine Room			

SUB-EQUIPMENT	APPROVED BY (Signature and title)
Refrigeration Machine	

Part or Component	No.	INSTRUCTIONS	Frequency (See notes below) T	D	1	2	4	12	26	52	Time
	13	The following items are applicable to <u>Absorption</u> machines only:									
		a. Check mechanical seals on pumps for leaks while machine is off.	x	x	x	x	x	x	x	x	1
		b. Check pressure in machine by use of temperature-pressure chart while machine is off		x	x	x	x	x	x	x	2
		c. Clean level probes in purge tank and reclaim solution. Follow book.			x	x	x	x	x	x	9
		d. Clean strainer and seal tank. Check for salt in seal water.			x	x	x	x	x	x	18
		e. Check specific gravity of solution in purge tank and in absorber for correct salt solution. Never add or remove water from machine as this is a critical change.					x	x	x	x	12
		f. Lubricate flow valves (quick opening).								x	5

NOTES: (1) T = Tour, D = Daily, 1 = 1 Week, 2 = 2 Weeks, etc.
(2) Frequency and Time are for one machine. Time is in minutes.

Table 8-1. Typical Routine Check List (page 4).

ROUTINE PREVENTIVE MAINTENANCE SERVICE CHECK LIST				NUMBER
SYSTEM AIR CONDITIONING	EQUIPMENT Refrig. Machine Room	ORIGINAL ISSUANCE DATE	REVISION(No.)	(Date)
SUB-EQUIPMENT Condenser and Chiller		APPROVED BY (Signature and title)		

Part or Component	No.	INSTRUCTIONS	Frequency (See notes below) T	D	1	2	4	12	26	52	Time
Condenser	1	Cleaning Condensers. If condenser water temperatures or pressures during the cooling season indicate possible clogging of tubes, take action as follows: a. Condenser With Removable Water Head- Remove heads and examine for scale. If evidence of scale, clean with rotary nylon brush. Never use wire brush.								R	300
		b. Condenser With Non-Removable Water Head - The tubes should be acid cleaned by qualified personnel or under contract.								R	90
	2	After cleaning and inspection, the condensers should be filled with treated water to prevent corrosion. The water level in the system should be kept low enough to prevent freezing.								R	30
Chiller	3	Chillers should be treated with chemicals introduced into the closed water circuit. Chillers need not be drained unless there is danger of freezing. If evidence of clogging, the same reference for cleaning condensers applies.								R	30

NOTES: (1) T = Tour, D = Daily, 1 = 1 Week, 2 = 2 Weeks, etc.
(2) Frequency and Time are one condenser or chiller. Time is in minutes.
R = As required.

Table 8-1. Typical Routine Check List (page 5).

ROUTINE PREVENTIVE MAINTENANCE SERVICE CHECK LIST				NUMBER
SYSTEM AIR CONDITIONING	EQUIPMENT Refrig. Machine Room	ORIGINAL ISSUANCE DATE	REVISION(No.)	(Date)
SUB-EQUIPMENT Pumps		APPROVED BY (Signature and title)		

Part or Component	No.	INSTRUCTIONS	Frequency T	D	(See notes below) 1	2	4	12	26	52	Time
Pumps	1	Check pumps for suction and discharge pressures. Also check motor and pump bearings by feel for over-heating.			x	x	x	x	x	x	2
	2	Check and adjust packing glands. (Water should drip slowly from packing gland, about one drop every two seconds when pump is off.) Keep adjusting nuts and bolts clean and lubricated. Note - Pumps with mechanical seals need no adjusting.			x	x	x	x	x	x	10
	3	Lubricate motor and pump bearings in accordance with manufacturer's recommendations.							x	x	20
	4	Check and observe operation of all pumps at the end of the cooling season to determine what parts may be needed to put each pump in proper operating condition during the off-season.								x	60
	5	Replace packing when adjustment has to be made too frequently or when packing gland is screwed in to the limit. Always check condition of shaft when the packing is removed and, if shaft sleeve is worn, it should be replaced before new packing is installed.								R	60

NOTES: (1) T = Tour, D = Daily, 1 = 1 Week, 2 = 2 Weeks, etc.
(2) Frequency and Time are for one pump. Time is in minutes. R = As required.

Table 8-1. Typical Routine Check List (page 6).

ROUTINE PREVENTIVE MAINTENANCE SERVICE CHECK LIST

ROUTINE PREVENTIVE MAINTENANCE SERVICE CHECK LIST			NUMBER	
SYSTEM AIR CONDITIONING	EQUIPMENT Refrig. Machine Room	ORIGINAL ISSUANCE DATE	REVISION(No.)	(Date)
SUB-EQUIPMENT Pneumatic Air Compressor for Controls		APPROVED BY (Signature and title)		

Part or Component	No.	INSTRUCTIONS	Frequency								Time
					(See notes below)						
			T	D	1	2	4	12	26	52	
	1	Check air pressure in supply tank and the pressure regulator adjustment in supply line for proper limits.			x	x	x	x	x	x	2
	2	Drain water from air tanks and supply line filters			x	x	x	x	x	x	2
	3	Check air compressor oil level, if low add oil.			x	x	x	x	x	x	1
	4	Check belt tension and alignment			x	x	x	x	x	x	1
	5	Test air relief valve per manufacturer's instructions					x	x	x	x	1
	6	Clean air compressor intake filter pads.					x	x	x	x	10
	7	Check all mounting screws and bolts for tightness.					x	x	x	x	3
	8	Check all connections for air leaks using a soap solution. If air compressor runs excessively, check for loss of air pressure at controls and throughout piping system.					x	x	x	x	5
	9	Change air compressor crankcase oil.								x	20
	10	Oil electric motor bearings per manufacturer's recommendation								R	2

NOTES: (1) T = Tour, D = Daily, 1 = 1 Week, 2 = 2 Weeks, etc.
(2) Frequency and Time are for one complete compressor. Time is in minutes.
R = As required.

Table 8-1. Typical Routine Check List (page 7).

ROUTINE PREVENTIVE MAINTENANCE SERVICE CHECK LIST

NUMBER				
SYSTEM: AIR CONDITIONING	EQUIPMENT: Outside Machine Room	ORIGINAL ISSUANCE DATE	REVISION (No.)	(Date)
SUB-EPUIPMENT: Cooling Tower		APPROVED BY (Signature and title)		

Part or Component	No.	INSTRUCTIONS	T	D	1	2	4	12	26	52	Time
					Frequency (See notes below)						
	1	Just prior to the beginning of the cooling season perform the following: a. Brush out tower.								x	30
		b. Where there are wood towers, caulk cracks and openings in tongue and groove joints using a soft mastic. Tighten all bolts. Paint any spots that have appeared since the shut-down painting.								x	30
		c. Clean and lubricate fan bearings.								x	15
		d. Install V-belts on fan assembly. Adjust to proper tension and alignment.								x	10
		e. Clean and lubricate electric motor bearings per manufacturer's recommendation.								x	20
		f. Operate fan motor and fan. Check motor and fan bearing for over-heating.								x	10
		g. Introduce chemicals for water treatment (condenser and chiller). Where you have an automatic feeder, adjust for proper feed.								x	40
		h. Fill system to proper basin level for your equipment operation. Check water level and float adjustment.								x	40
		i. Start-up cooling system and adjust bleed-rate as per your chemical supplier's recommendation.								x	30

NOTES: (1) T = Tour, D = Daily, 1 = 1 Week, 2 = 2 Weeks, etc.
(2) Frequency and Time are for one complete tower. Time is in minutes.

Table 8-1. Typical Routine Check List (page 8).

ROUTINE PREVENTIVE MAINTENANCE SERVICE CHECK LIST — NUMBER

SYSTEM	EQUIPMENT	ORIGINAL ISSUANCE DATE	REVISION(No.)	(Date)
AIR CONDITIONING	Outside Machine Room			

SUB-EQUIPMENT	APPROVED BY (Signature and title)
Cooling Tower	

Part or Component	No.	INSTRUCTIONS	Frequency (See notes below) T	D	1	2	4	12	26	52	Time
	2	During cooling season check tower as follows:									
		a. Check fan by visual inspection of V-belt conditions and alignment. Check fan and motor for unusual noise or vibration.			x	x	x	x	x	x	5
		b. Check basin water level and float.			x	x	x	x	x	x	1
		c. Check bleed-rate. Adjust as required.			x	x	x	x	x	x	10
		d. Observe nozzle spray (and screens if used) for clogging.			x	x	x	x	x	x	2
		e. Check water treatment.			x	x	x	x	x	x	5
		f. Clean intake strainer.				x	x	x	x	x	15
		g. Make fresh water float valve adjustment for proper water level and overflow control, when required.				x	x	x	x	x	10
		h. Check tower basin accumulation of mud, silt, slime, algae, etc. Drain, hose out, and fill as required.				x	x	x	x	x	60
		i. Lubricate fan bearings as required.					x	x	x	x	10
		j. Check shaft couplings, if so equipped, and oil level in gear unit as manufacturer recommends.					x	x	x	x	5
		k. Check steel casing and metal parts of tower for rust and corrosion.					x	x	x	x	5
		l. Perform any repairs which are urgent and schedule remainder for performance when tower is shut down at end of season.								R	

NOTES: (1) T = Tour, D = Daily, 1 = 1 Week, 2 = 2 Weeks, etc.
(2) Frequency and Time are for one complete tower. Time is in minutes.
R = As Required.

Table 8-1. Typical Routine Check List (page 9).

ROUTINE PREVENTIVE MAINTENANCE SERVICE CHECK LIST			NUMBER
SYSTEM AIR CONDITIONING	EQUIPMENT Outside Machine Room	ORIGINAL ISSUANCE DATE	REVISION(No.) (Date)
SUB-EQUIPMENT Cooling Tower		APPROVED BY (Signature and title)	

Part or component	No.	INSTRUCTIONS	Frequency (See notes below) T	D	1	2	4	12	26	52	Time
	3	At end of cooling season perform following: a. Clean and flush out tower, leaving drain open.								x	75
		b. Drain only those parts of tower system likely to freeze. If necessary to drain entire system, leave all valves open. Clean drift eliminators, fill deck, louvres, and spray heads.								x	30
		c. Remove V-belts from fan assembly and store out of the weather.								x	15
		d. Have repairs made as required during off season for placing tower in operation when needed.								R	
		e. When tower is completely dry, clean all surfaces that need painting. Apply a primer having a chromate base. Then apply either a marine type paint, or paint as recommended for your type of tower. Note - The frequency for painting may be once in two or three years, depending on local conditions. Redwood towers with corrigated asbestos casing require no painting.								R	

NOTES: (1) T = Tour, D = Daily, 1 = 1 Week, 2 = 2 Weeks, etc.
(2) Frequency and Time are for one complete tower. Time is in minutes.
R = As Required.

Table 8-1. Typical Routine Check List (page 10).

ROUTINE PREVENTIVE MAINTENANCE SERVICE CHECK LIST				NUMBER
SYSTEM AIR CONDITIONING	EQUIPMENT Outside Machine Room	ORIGINAL ISSUANCE DATE	REVISION(No.)	(Date)
SUB-EQUIPMENT Air Handlers		APPROVED BY (Signature and title)		

Part or Component	No.	INSTRUCTIONS	Frequency (See notes below)								Time
			T	D	1	2	4	12	26	52	
	1	Record data on Service Card. Check bearing temperatures of motor and fan by feel			x	x	x	x	x	x	4
	2	Check outside air-intake screens; clean if necessary					x	x	x	x	4
	3	Lubricate fan bearings as recommended.					x	x	x	x	2
	4	Check belts for proper tension, and pulleys for proper adjustment.					x	x	x	x	3
	5	Check filters. Replace or clean when necessary, depending on type. Look for tears.					x	x	x	x	20
	6	Check coils. If not clean, they must be air cleaned, flush cleaned or steam cleaned depending upon condition.					x	x	x	x	15
	7	Lubricate electric motor bearing as recommended by manufacturer.								R	3
	8	Check operation of dampers, automatic valves, and linkages. Make certain fresh-air intake damper is not open beyond the minimum position unless called for.							x	x	15
	9	Check fan shaft for end play, vibration, and bearing wear.							x	x	10
	10	Clean and flush condensate pan and its drain lines.								x	60
	11	Check fan blades for looseness and dirty condition. Clean if necessary.								x	10
	12	Check all junction box and contactor connections on fan motors for tightness by qualified personnel. Check condition of contactor points. Have replacements made as needed.								x	30
	13	Where possible, observe entire duct system for loose, or missing insulation, openings, caved-in duct, fire dampers, etc., especially on new installations.								x	15
	14	Clean all damper linkages with a solvent. Lubricate with graphite and penetrating oil.								x	60

NOTES: (1) T = Tour, D = Daily, 1 = 1 Week, 2 = 2 Weeks, etc.
(2) Frequency and Time are for one complete air handler. Time is in minutes.
R = As Required.

Table 8-1. Typical Routine Check List (page 11).

ROUTINE PREVENTIVE MAINTENANCE SERVICE CHECK LIST

NUMBER

SYSTEM	EQUIPMENT	ORIGINAL ISSUANCE DATE	REVISION(No.)	(Date)
AIR CONDITIONING	Outside Machine Room			

SUB-EQUIPMENT: Controls

APPROVED BY (Signature and title)

Part or Component	No.	INSTRUCTIONS	Frequency (See notes below) T	D	1	2	4	12	26	52	Time
	1	Check room thermostat settings. Where have pneumatic thermostats, use air gauge tester.							x	x	5
	2	Check operation of mixing valves, damper motors, etc. Clean and adjust if necessary.							x	x	10
	3	Observe operation of entire control system. Check for anything not operating in a normal manner. If qualified personnel not available, report to the region.							x	x	180

NOTES: (1) T = Tour, D = Daily, 1 = 1 Week, 2 = 2 Weeks, etc.
(2) Frequency and Time are for each unit indicated. Time is in minutes.

so, the time shown for each item represents only the time required to perform that task once. It does not reflect the number of men required for that task. For the larger installations, different frequencies and times are necessary for certain items in recognition of the type and size of installation. Wherever warranted, adjust frequencies of inspections to meet local operating conditions.

Maintenance Program

The extent of any maintenance program depends mainly on the amount of capital invested in equipment, the number and skill of maintenance personnel, and the importance of eliminating equipment down-time. Where down-time is of prime importance, extensive maintenance programs are justified. However, in the case of large commercial buildings, a good, modified maintenance program can be set up for air conditioning equipment by using a check list similar to the one provided here. With such a check list used as a base (with the items numbered and identified), a fairly uncomplicated program can be developed. The program should be set up and administered under the direction of the chief engineer or other comparably qualified member of the maintenance force responsible for the operation of the equipment.

Routine Inspection Tour — Assign each maintenance mechanic an inspection tour for each day.

Route Sheet — The maintenance mechanic will take with him a previously prepared route sheet listing the equipment to be visited on his tour for the day. The equipment and items to be covered on the route sheet will tie-in directly with the items on the basic check list. The mechanic should be trained in the procedures to follow and be familiar with the steps in carrying out the inspection for each item. For example, where lubrication is involved, the type and grade of grease, as well as the amount and frequency recommended by the manufacturer should be spelled out; where damper linkages are concerned, the manufacturer's lubrication instructions for the damper rods and bushings should be spelled out, etc. The trained mechanic should automatically notice things (during his tour) which could cause shut-downs and costly major repairs if allowed to continue. These failings should be reported immediately so that they can promptly be given the required attention.

How to Inspect — In carrying out the items spelled out in the "Route Sheet" the mechanic should watch for the following:

1. Notice any abnormal changes in the meter, gauge, and thermostat readings, and ascertain the reason for the change.
2. Keep an eye on the condition of all fittings, caps, and flanges on the compressor(s). If any show signs of being oily, they should be checked with soap bubbles or a Halide Torch for refrigerant leaks.
3. He should be familiar with the sound of both the compressor and motor when they are operating properly, so that he can distinguish any abnormal sounds immediately and ascertain the cause.
4. Presuming knowledge of the function and operation of the various controls, he should observe if they are operating properly.

Contract versus Breakdown Service — When qualified maintenance personnel is not available, the decision must be made whether to enter into an annual service contract (referred to in Chapter 10) with an air conditioning concern or to operate on a breakdown basis.

Annual Service Contract — This type of contract has the advantage of assuring a continuity of operation through preventive maintenance with the contractor available on short notice. It also provides opportunity for maintenance personnel to learn preventive maintenance by observing the contractor's men perform. However, the annual contract service is usually expensive, depending on the size and complexity of the installation.

Breakdown Service — Under this arrangement, it is possible to get by for a long period with little or no expense, but it is a gamble for just when it is needed the most, service may not be available for a long period or might be necessary just when time-and-a-half or double-time rates would have to be paid. It is practical with a new installation, after the initial shake-down and warranty period have elapsed, but not so after a year or two without qualified maintenance personnel, when it is definitely a gamble.

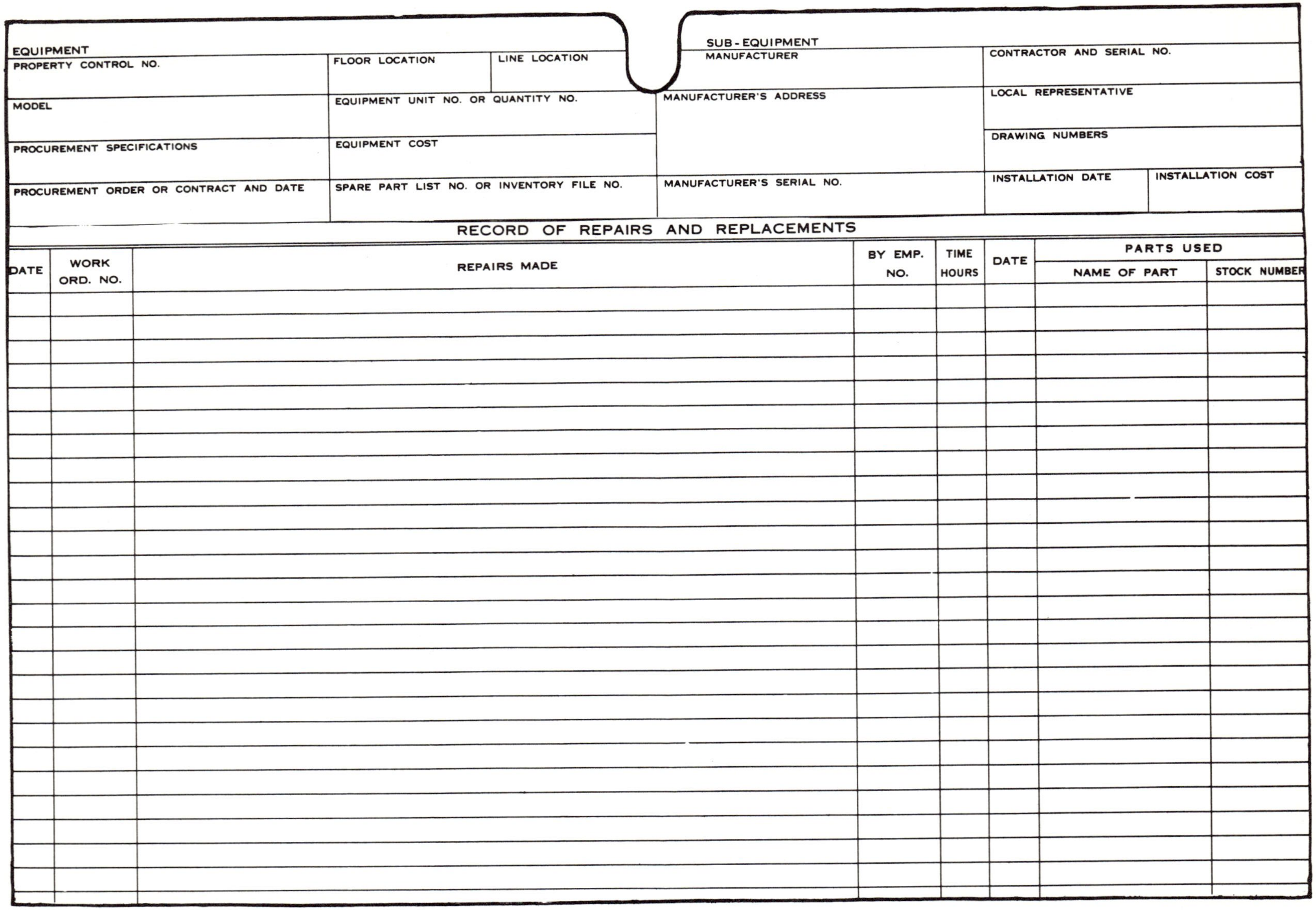

EQUIPMENT			SUB-EQUIPMENT		
PROPERTY CONTROL NO.	FLOOR LOCATION	LINE LOCATION	MANUFACTURER	CONTRACTOR AND SERIAL NO.	
MODEL	EQUIPMENT UNIT NO. OR QUANTITY NO.		MANUFACTURER'S ADDRESS	LOCAL REPRESENTATIVE	
PROCUREMENT SPECIFICATIONS	EQUIPMENT COST			DRAWING NUMBERS	
PROCUREMENT ORDER OR CONTRACT AND DATE	SPARE PART LIST NO. OR INVENTORY FILE NO.		MANUFACTURER'S SERIAL NO.	INSTALLATION DATE	INSTALLATION COST

RECORD OF REPAIRS AND REPLACEMENTS

DATE	WORK ORD. NO.	REPAIRS MADE	BY EMP. NO.	TIME HOURS	DATE	PARTS USED	
						NAME OF PART	STOCK NUMBER

Exhibit 8-A. Equipment record card.

Equipment Record Card — A record card should be prepared for each piece of equipment. The card should show all pertinent data relating to the equipment including dates inspected, malfunctions, repairs and their costs, etc. Exhibit 8-A shows such a card. A record such as this offers an excellent means for comparative performance evaluation.

CHAPTER 9

Economical Operation

General

The material herein provides specific means whereby air-conditioning costs can be reduced in all medium-to-large buildings. While the subject "Load Control" requires some skill on the part of operating personnel, most of the dollar-saving practices outlined can be performed by anyone and with very little effort result in worthwhile dollar savings.

Electricity Economics

Background — Next to cleaning and maintenance, electricity is the greatest single expense in the operation of a large building. Through proper control of its use, costs can be reduced and held to a minimum. The program outlined goes beyond the usual practices, such as turning off lights when not needed, using small wattage bulbs, etc. It concerns itself with the "choice" of the most advantageous rate and "load control."

Choosing the Lowest Rate — To begin with, all utility companies have more than one rate that can be applied to a building. A choice is made when the building is new and, thereafter, it is the *responsibility* of the customer to see that he is billed at the lowest rate possible. What may have been the lowest rate when the building was new, may not be so several years later, since the electrical needs of a building vary with the kind and extent of equipment used by tenants and also with the type of occupancy. Changes such as these not only affect the consumption of electricity (KWH) but also affect the KW Demand and resultant load factor.

Thus, to insure lowest billing, the electric consumption and load conditions of a building should be checked periodically against available rates.

Rate Structure — (See Exhibit 9-A).

Types of Charges — The electric rate applied to large buildings usually incorporates two separate charges in the rate: (1) a Demand Charge based on Kilowatts (KW), and (2) an Energy Charge based on Kilowatt Hours (KWH).

Kilowatt Demand (KW)

a. *Description* — This part of the rate is actually a "rate of use" or "readiness to serve" charge imposed by the company in recognition of the fact that they furnish and maintain generating and supply equipment of sufficient capacity to meet a consumer's "demand" at all times. It is not a measure of the *amount* of electricity used, but a measure of the *rate* at which it is being supplied by the company over a specified period of time. The rate will specify this period as being 15, 20, or 30 minutes. It makes no difference when the maximum demand occurs. It could occur on the last day of the billing period and still would be applied to the bill for the month. In some cases it is applied to bills for the following 11 months even though smaller demands are registered during these subsequent months.
b. *Example* — Assume the rate provides that the maximum demand be measured over a 15-minute period. Assume a condition when no air conditioning equipment is on the line and the maximum demand so far during the month for the building load only (lights, elevators, etc.), is 500 KW. If one of two 450-ton refrigeration machines (with a total load of 500 KW) is cut-in and runs for at least 15 minutes, the maximum demand would

LARGE COMMERCIAL SCHEDULE
ALTERNATING CURRENT

Effective February 15, 1960

Applicable to any commercial consumer using Alternating Current and having a demand equal to or in excess of 20 KW.

1. RATE

(a) Kilowatt Demand Charge

For the first	20 KW demand per month per KW	$2.32
For the next	30 KW demand per month per KW	1.69
For all excess over	50 KW demand per month per KW	1.18

(b) Kilowatt-Hour Charge

For the first	40,000 KWH used per month per KWH	$ .0161
For the next	60,000 KWH used per month per KWH	.012
For all excess over	100,000 KWH used per month per KWH	.0091

(c) Fuel Adjustment

If the weighted average cost of fuel for electric generation for the calendar month preceding the billing date, ascertained by the Company and filed monthly with the Public Utilities Commission of Ohio, is above or below $6.80 per ton of 2,000 lbs of coal or its equivalent in heat content of other fuel, the energy charges specified above shall be increased or decreased $.000005 per KWH for each full one cent by which the cost of coal is above or below $6.80 per ton.

2. APPLICATION

Assume Max. Demand = 2,626 KW
Assume Energy = 1,058,400 KWH

DEMAND CHARGE:

20 KW	@	$2.32............	$ 46.40	
30 KW	@	$1.69............	50.70	
2,576 KW	@	$1.18............	3,039.68	
2,626 KW				$ 3,137

ENERGY CHARGE:

40,000 KWH	@	$0.0161..........	$ 644.00	
60,000 KWH	@	$0.0120..........	720.00	
958,400 KWH	@	$0.0091..........	8,721.44	9,085
1,058,400 KWH			TOTAL	$12,222

Exhibit 9-A. Typical electric demand rate and application.

be 500 + 500 = 1,000 KW. Now assume the second machine is also cut-in during the same month and both machines are on the line during the same 15 minutes. The Demand would then be: 500 + 1,000 = 1,500 KW. It can readily be seen, therefore, that the number of machines on the line at any one time has a direct bearing on the maximum demand; and the larger the machine, the greater the effect.

c. *Billing* — Using the same demand rate shown in Exhibit 9-A, the demand charge for the conditions outlined in (b) would be as follows:

No Air Conditioning:	
500 KW Dem.	$ 628.10
One Refrig. Mach. on line:	
1,000 KW Dem.	$1,218.10
Two Refrig. Machs. on line:	
1,500 KW Dem.	$1,808.10

Kilowatt Hours of Energy (KWH)

a. *Description* — This part of the rate is a measurement of the "amount" of electricity used during the month and thus differs from the KW Demand which is a measurement of the "rate of use."

b. *Example* — In the example reference was to a *maximum* demand of 500 KW for the building load only. Assume a month in November when no air conditioning is needed. Remember 500 KW was the maximum registered during any 15-minute period during any day of that month, whereas the actual demand could have varied between 275 and 500. Say the average was 350 KW. Using an average demand of 350 KW and 30 days for the month of November, the total energy used in KWH for the 720 hours in that month would be:

350 KW × 720 Hours = 252,000 KWH

c. *Billing* — Using the same KWH charge shown in Exhibit 9-A the energy charge for the 252,000 KWH would be:

40,000 KWH @ $.0161	$ 644.00
60,000 KWH @ .012	720.00
152,000 KWH @ .0091	1,383.20
252,000 KWH	$2,747.20

d. The total charge for the month of November based on these examples and using the rate shown in Exhibit 9-A, would be:

Demand: 500 KW	$ 628.10
Energy: 252,000 KWH	$2,747.20
Total Charge:	$3,375.30

Load Control Charts

Managing — From the examples above, it is apparent that, through control over the rate of use of electricity (Demand), we can exercise direct control over the cost of electricity, particularly in large buildings.

Recording Demand Meter — The first step is to request the utility company to install a recording demand meter, if one is not already installed. The next thing is to learn to "read" the demand chart of the meter and utilize it to advantage.

Plotting Dollar Savings through Load Control Charts (Combination Office-Industrial Building)

a. *Summer Load Chart* — Exhibit 9-B shows the maximum demands taken from a company's demand meter chart for the month of July and plotted for each hour of the day. The Demand Charge *for the month* was based on the 4,680 KW *peak* registered between 11:30 A.M. and noon on July 24.

b. *Winter Load Chart* — The maximum demands for the month of November for the same building, with no air conditioning load, is shown in Exhibit 9-C. This can be considered as representing the general building load only — that is, all lights and motors other than those connected to the cooling system. Note that the range during this month was from 2,220 KW at midnight to a maximum of 2,820 KW at 10:30 A.M. At $1.40 per KW, the Demand Charge for this month would be $3,948.00

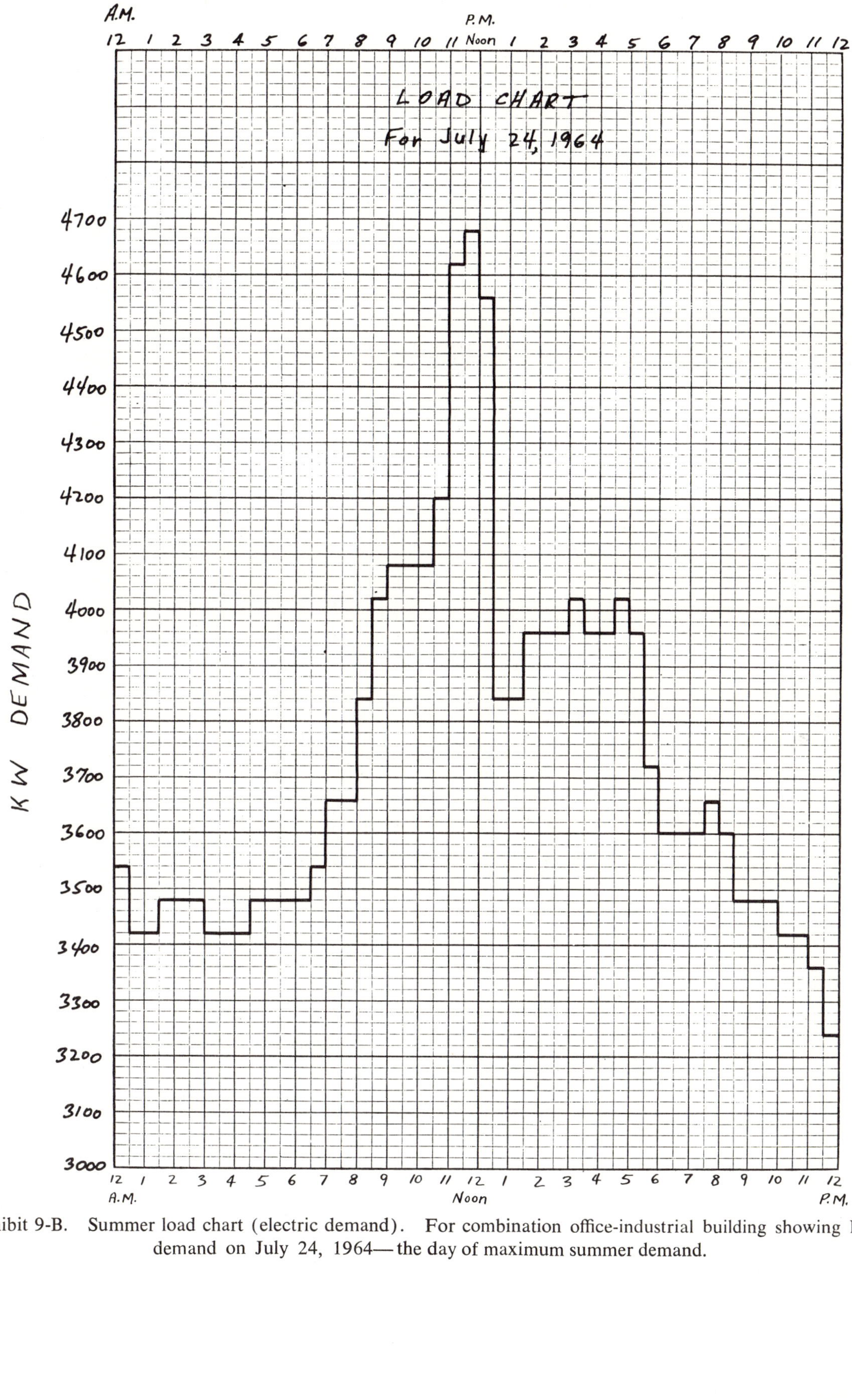

Exhibit 9-B. Summer load chart (electric demand). For combination office-industrial building showing KW demand on July 24, 1964—the day of maximum summer demand.

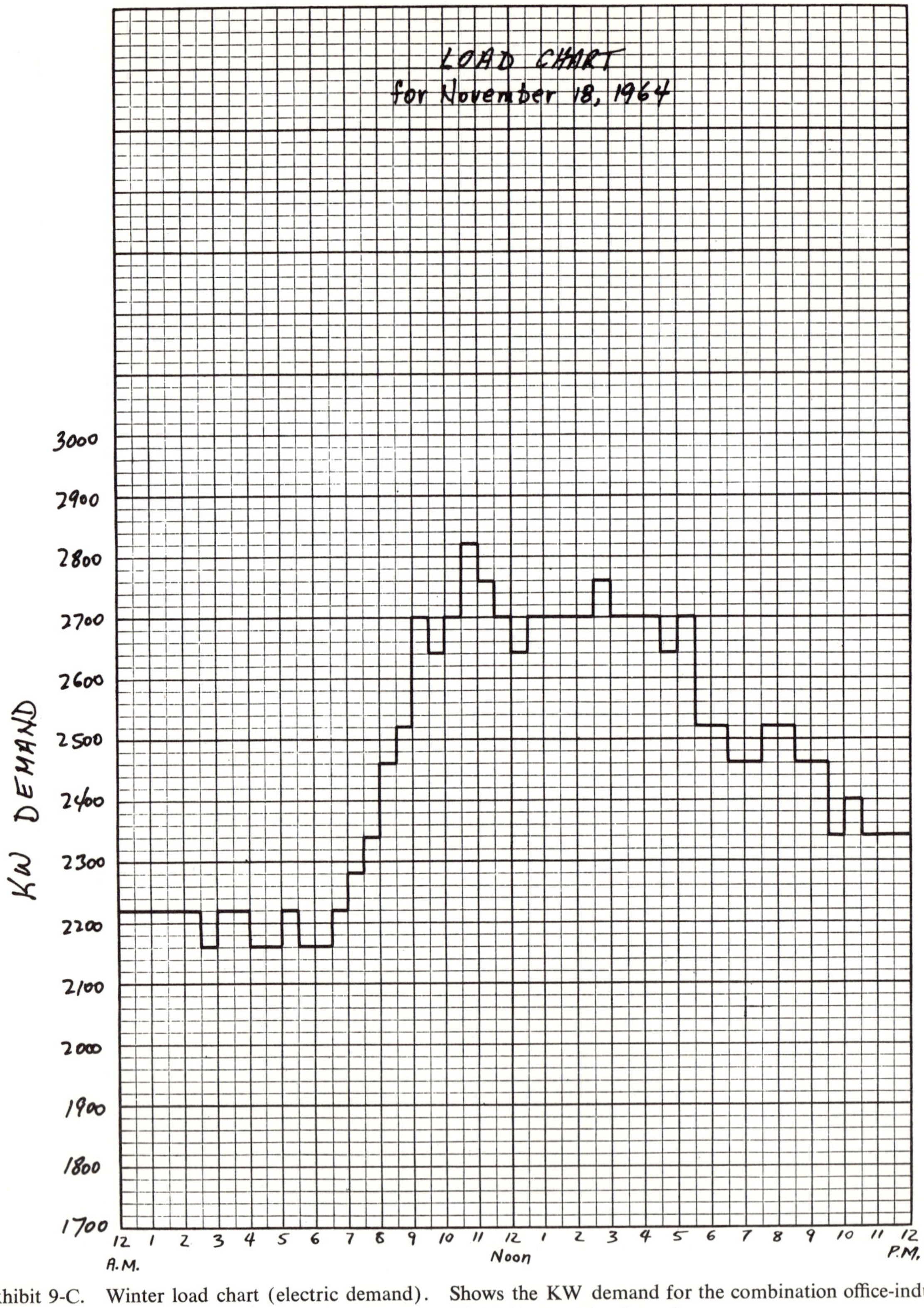

Exhibit 9-C. Winter load chart (electric demand). Shows the KW demand for the combination office-industrial building on November 18, 1964—a peak winter day.

compared to $6,552.00 for a demand of 4,680 KW; a difference of $2,600.00.

c. *Combined Summer and Winter Chart* — Exhibit 9-D shows the maximum demands for two months plotted on one chart. This combined chart permits comparison between the magnitude of the demands for these two months *and* the corresponding time of day when the demands occurred. (In order to get the two months on one chart and use a workable scale, it was necessary to break the chart at the 2,875 KW level and omit the 3,000 and 3,125 lines.) It can readily be seen from this combined chart that the time of day to watch in this building is between 10:30 A.M. and 12:30 P.M.

Large Commercial Office Building — The load chart for this building (Exhibit 9-E), compares the electric load with and without air conditioning in operation. It will be noted that, in this building, the peak demand occurs between 1:00 and 4:00 P.M., whereas the peak time for the Office-Industrial building is registered at noon.

Large Apartment Building — For this building, the load chart (Exhibit 9-F), also compares the electric load with and without air conditioning in operation. It will be noted that the peak demand in this building occurred between 9:00 and 11:30 P.M.

Evaluation — The value of preparing a load chart for a particular building can be seen from the foregoing. In the case of the combination office-industrial building, it was indicated that the time of day to initiate load control was at noon, for the large commercial building it was between 1:00 and 4:00 P.M., and for the apartment building it was between 9:00 and 11:00 P.M. Each building presents its own particular conditions of operation.

In the case of the apartment building in Exhibit 9-F, since the peak demand occurs late at night, it should be an easy matter to reduce the demand (provided the humidity is not excessive). This is particularly true where the building is equipped with two refrigeration machines and it is the customary practice to run *both* machines all night long.

Load Factor

The Load Factor is an excellent indicator of the value of good load control. It is represented by the following ratio:

(part 1)

$$\frac{\text{Av. Demand}}{\text{Max. Demand}} =$$

(part 2)

$$\frac{\text{Av. Demand} \times (24 \text{ hrs.} \times \text{No. of days in Billing Period})}{\text{Max. Demand} \times (24 \text{ hrs.} \times \text{No. of days in Billing Period})} =$$

(part 3)

$$\frac{\text{Total KWH for the month}}{\text{Max. Demand} \times \text{No. of hrs. in Billing Period}}$$

To illustrate this, let us take the Summer Load Chart (Exhibit 9-B). The 7-day chart from which the peak maximum demand of 4,680 KW was obtained on July 24th, shows the following maximum demands registered for the 4 previous days, and also those for the following 2 days:

7/20	4,080 KW
7/21	4,080 KW
7/22	4,200 KW
7/23	4,200 KW
7/24	4,680 KW*
7/25	3,360 KW
7/26	3,600 KW

* Peak Demand

If we omitted the Peak Maximum Demand registered on July 24th, the Average Maximum Demands of the other 6 days would only be 3,910 KW.

The maximum Dry-Bulb temperature for this day was only 80° F and the minimum was as low as 70° F. The Wet-Bulb temperature is not known, but even if it were fairly high, it would not explain so large a difference from the average of 3,910 KW. We will now see how the Load Factor would have reflected this type of operation.

Company records for the month show: Demand Billing based on 4,680 KW; total consumption for the month — 2,669,000 KWH; and number of days in the Billing Period — 32 days. (The 32 days results in a total of 768 hours in the Billing Period.) Using these figures in *Ratio, part 3,* re-

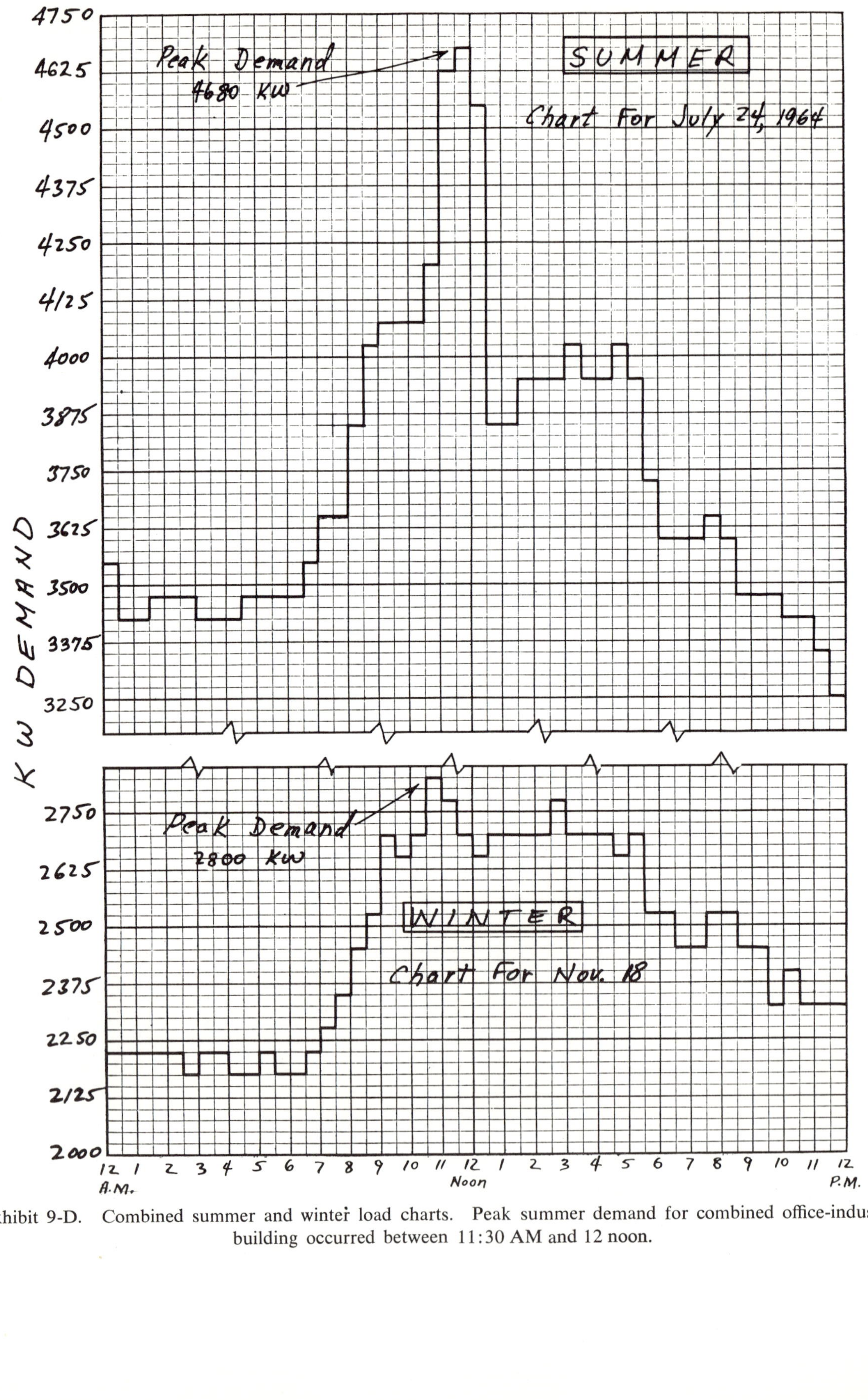

Exhibit 9-D. Combined summer and winter load charts. Peak summer demand for combined office-industrial building occurred between 11:30 AM and 12 noon.

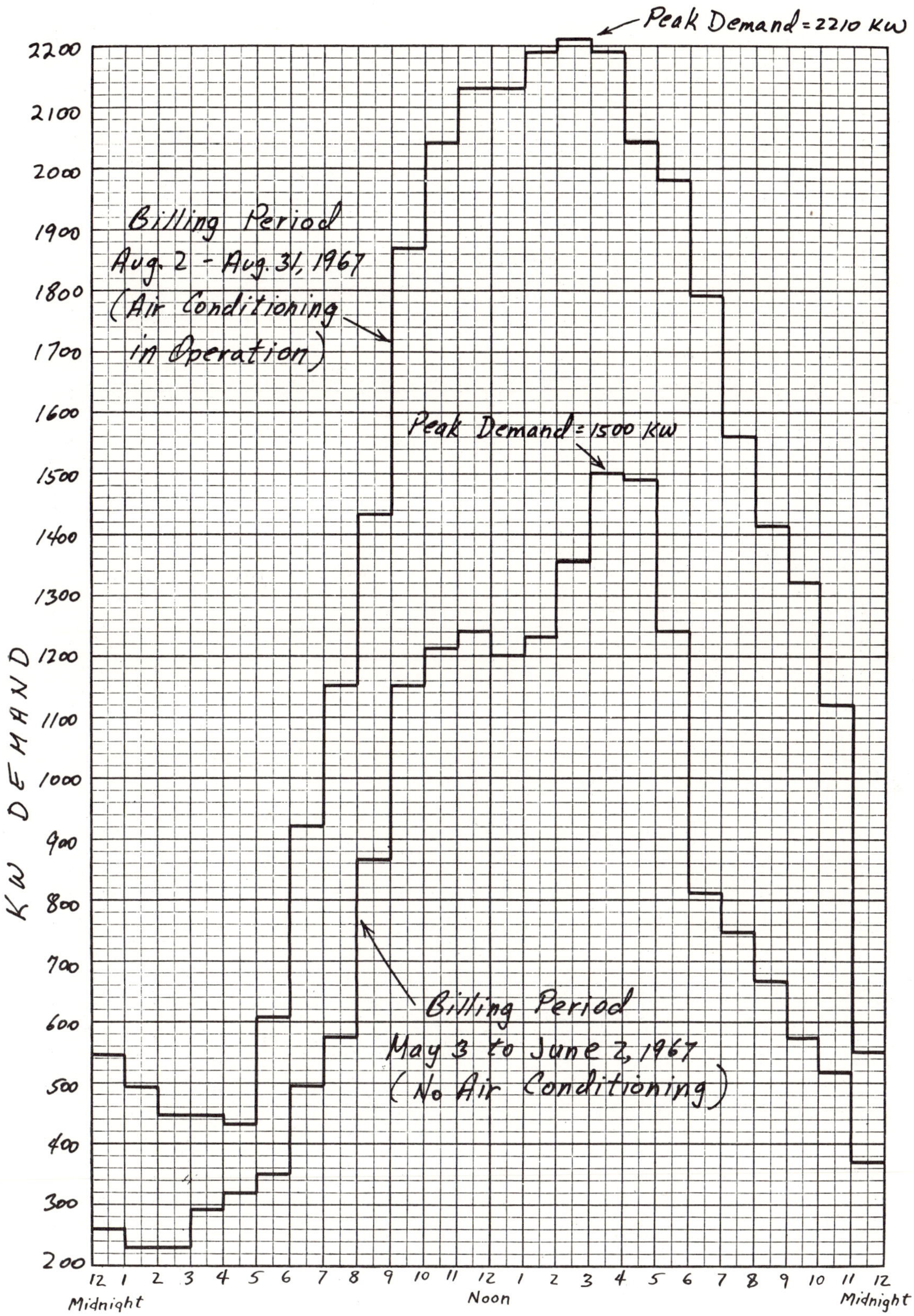

Exhibit 9-E. Load chart for large office building.

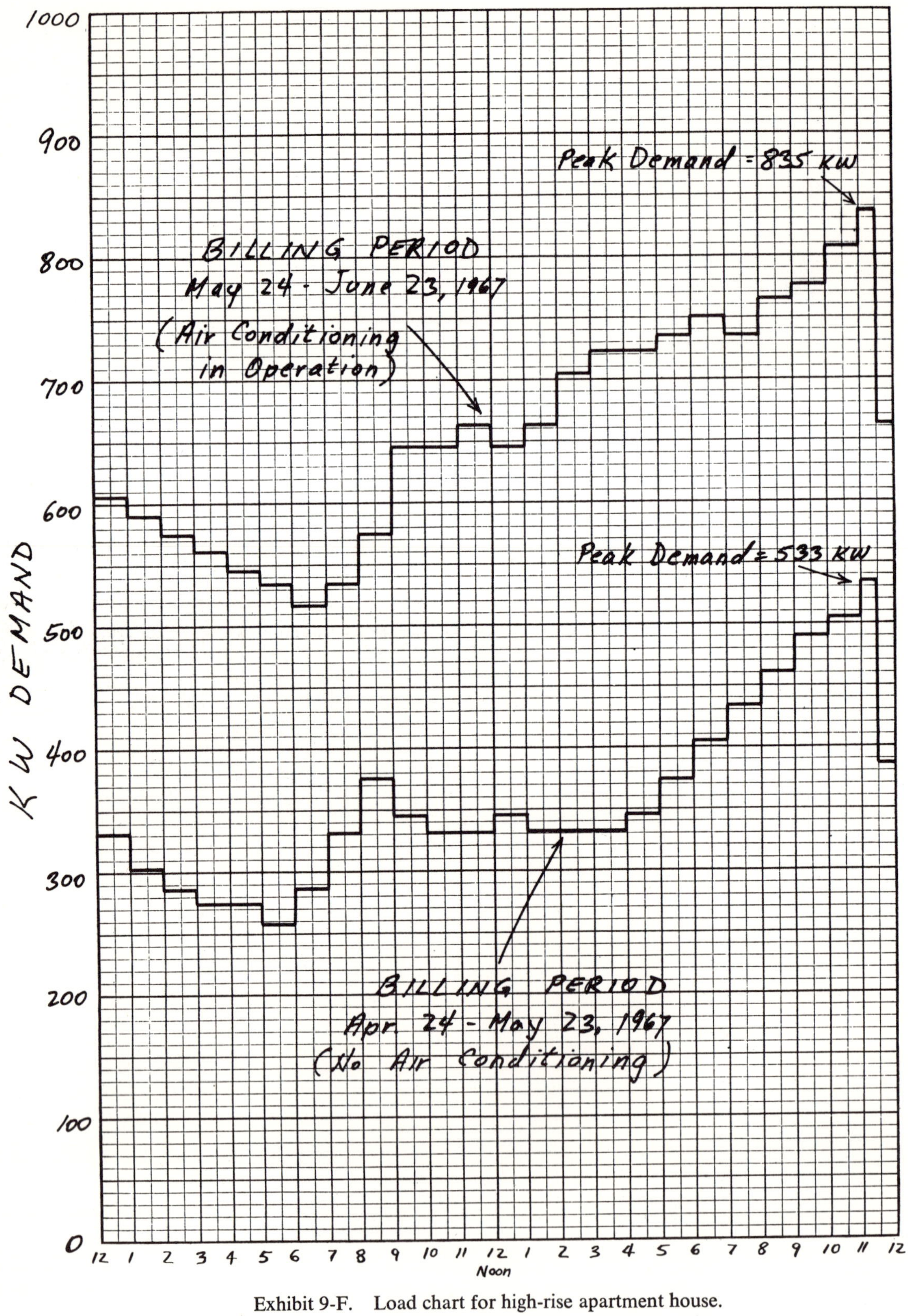

Exhibit 9-F. Load chart for high-rise apartment house.

sults in a Load Factor as follows:

$$\text{Load Factor} = \frac{2,669,000}{4,680 \times 768} = 74.2\%$$

Assume that, through Load Control, the maximum demand had been held to 4,200 KW. In addition to a reduction in Demand of 468 KW, there would have been a reduction of 18,600 KWH for the month. Again using *Ratio, part 3,* the new Load Factor would be:

$$\text{Load Factor} = \frac{2,650,400}{4,200 \times 768} = 82.2\%$$

Savings for the month at the rate shown in Exhibit 9-A would be:

Demand:	468 KW	@ $1.18	$553.00
Energy:	18,600 KWH	@ $0.0091	169.00
		Total:	$722.00

Total savings for a 5-month cooling season = $3,610.00

Corrective Action — Suppose the building represented by the load chart, Exhibit 9-B, had two 1,200-ton refrigeration machines and was normally operated with a leaving chilled-water temperature of 45° F, and only one of the two machines was operated in the early morning hours.

The first consideration would be to see which large motors could be taken off the line between the hours of 8:00 A.M. and 12 noon. It is possible that certain elevators, large air-compressor motors, fire storage-tank pumps, artesian-well pumps, sewage pumps, battery chargers, etc., could be kept off the line during this time. It is also possible that sub-cooling could be practiced during the period 7:00 A.M. to 10:30 A.M., by reducing the leaving chilled-water to 40° or 41° F, utilizing only one machine. Thus, the sub-cooled circulating chilled-water would act as a "fly-wheel" in carrying the building over the peak period from 11:00 A.M. to 12:30 P.M., at a Peak Demand of possibly even less than 4,200 KW, instead of using both machines and registering a Peak Demand of 4,680 KW. Where the refrigeration plant consisted of only one machine, the machine could be shut down after sub-cooling for the required length of time and only the chilled-water circulating pump operated to produce the "fly-wheel" effect. The main objective is to keep as many large machines as possible off the line during the time of peak *Building* demand.

Demand Period — When the electric rate is based on a 20-minute demand period, and it is not possible to shut off one or both machines for the entire length of time required, one machine may be shut off after operating 10 or 15 minutes and then cut back on, *after* the 20-minute peak demand period has passed, thus reducing the peak demand by ⅓ or ⅔ of the air conditioning load. A 30-minute demand period offers even larger possibilities of control.

Start-Up — When starting-up a plant where there are two or more machines, never start-up both machines at the same time. Wait at least the "Demand Period." As stated previously, this period may be 15, 20, or 30 minutes. It can also be ascertained by watching the demand-meter recording pen as it reaches its peak and returns to zero.

> CAUTION: In practicing sub-cooling, it would be necessary to exercise caution in not sub-cooling to the point where the freeze protection control would cause the compressor to cut-out. The operator would also have to know under what conditions of Wet-Bulb temperatures it could be done. But the opportunity for worthwhile savings is there to be taken advantage of by an experienced operator.

Beginning of Cooling Season — Assume that a building has: (1) a 1,000 KW refrigeration machine, (2) the maximum demand recorded for the month, from April 1st through April 27th, was 250 KW based on a 15-minute period, and (3) the meter is read on the 28th of the month at 2:15 P.M. If the 1,000 KW refrigeration machine had been cut on the line, on the 28th day at 2 P.M., the maximum demand could have been 1,250 KW or more. At $1.40 per KW demand, the cost of providing air-conditioning for these 15 minutes could be $1,400, whereas if the system had been cut-on at 2:20 P.M. (*after* the meter was read), a saving of $1,400 would have been realized on the April billing. Electric meters are usually read by utility companies on a regular schedule set up months in advance. This reading date should be ascertained and, if possible, even the time of day it is read.

Thus, when it is time to cut-on the air conditioning system, it should be scheduled, where possible, *after* the meter has been read for the month. *Remember, "heat is a must," whereas cooling represents comfort only.* When the need for air conditioning is such that it is not possible to hold off until the scheduled meter reading date, try to arrange with the utility company to read the meter at an earlier date. This is a must where the rate specifies that the maximum demand shall apply for the subsequent 11 months.

Load Chart Illustration

Exhibit 9-G is an actual Load Chart plotted for the first month of air conditioning for a large apartment house. It is an excellent example of the effect on the demand charge. If the plant had not been turned on May 18th and instead, was turned on May 23rd just after the meter was read, the saving for this month would have been $275.00; the difference between $1,213 and $938. If this had been a large commercial building, you can imagine what the saving would represent. The chart is also a good example of load control, as it shows that the cooling plant was cut-in during a low point in the building load. If it could have been turned on between 2 A.M. and 7 A.M. it would have been even better. What would have happened had the plant been cut-in between 8 P.M. and 9 P.M.? The peak or billing demand for this month would have been 950 KW and the demand charge would be $1,641.00 compared to the actual billing demand of $1,213.00. This is a difference of $428.00 bearing out the importance of knowing exactly when the maximum building load occurs (other than air conditioning), as explained previously under "Load Control."

> NOTE: The actual Demand Charge for this particular building is based on the following Demand Rate:
>
> | | | |
> |---|---|---|
> | First | 25 KW Demand | No charge |
> | Next | 75 KW Demand | @ $2.05 per KW |
> | Over | 100 KW Demand | @ $1.75 per KW |

Spring Check — In the spring season when it is necessary to operate and check the refrigeration plant prior to the beginning of the cooling season, the utility company should be so notified. The company will recognize the fact that this is merely a test period and may adjust the demand meter accordingly, prior to operation of the plant. In the case of larger buildings or complexes, the savings for this one month's bill could amount to thousands of dollars.

End of Cooling Season — The same instructions for the beginning of the cooling season apply for cutting-off the air conditioning at the end of the cooling season with the following adaptation:

> Assume the meter is read at 2:00 P.M. on Sept. 25th and the plant is shut down for the season at 2:15 P.M. The bill for the month of October would include the maximum demand charge registered during those first 15 minutes even though the plant was not operated again that month.

Water Bill Dollar Savings

Sewage Charge Background — The sewage charge in connection with water bills is usually based on the amount of water registered on the water meter. In other words, it is assumed that all water used in a building is discharged to the sewer. For those buildings having 100 tons of air conditioning and over, an appreciable saving can be made in water bills where a municipality will recognize the fact that only about 16 per cent of the make-up water going to the cooling tower is piped to the sewer; the balance of the water (about 84 per cent) is lost through evaporation and windage.

Measurement — The water lost to evaporation and windage can be measured by installing two water meters — one in the make-up line and one in the bleed line and then taking the difference between the two readings. This amount of water is deducted from the total amount of water measured by the building meter and the sewage charge is based on this remainder.

Cost — Each tower would require two meters at an average cost of approximately $40.00 per meter. They can be installed by the maintenance

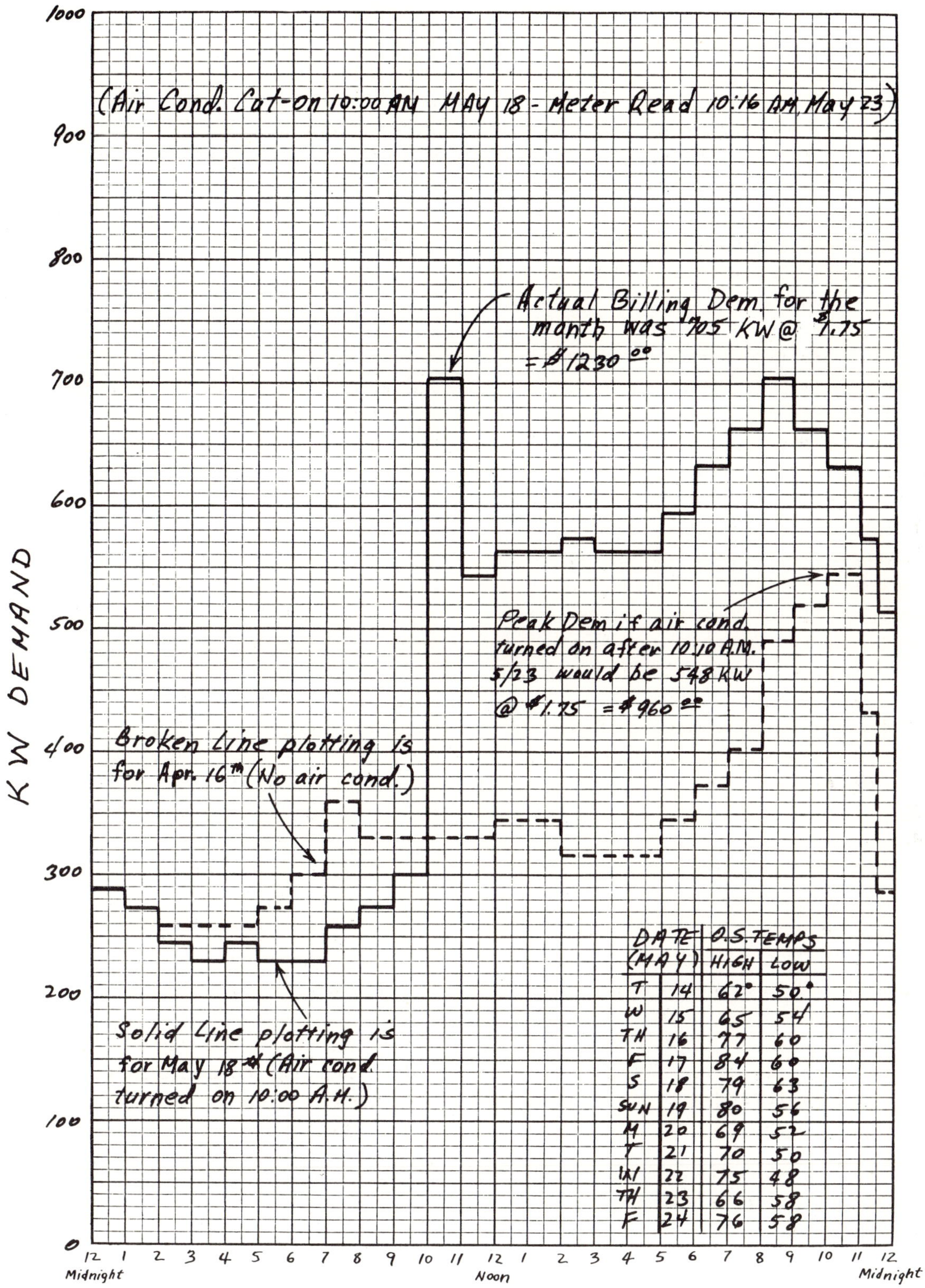

Exhibit 9-G. Load chart for apartment house. (Air conditioning versus no air conditioning.)

personnel at the larger buildings and in some of the smaller and medium-size buildings.

Alternative — Another method (if agreeable to the utility company) is to use only the one meter on the make-up line and have them read the meter at the begining and end of the cooling season and not apply any sewage charge for the amount of water measured by this meter. Some utility companies permit this practice, since the bleed is a relatively small amount compared to the total make-up water entering the tower.

Miscellaneous Dollar Savings

Supervision — Assign responsibility for control to area supervisors. Watch out for lunch periods, after-work hours, non-occupied areas, and weekends. Instruct guards to turn off lights during their rounds wherever not needed, and also motors that have been left running.

> NOTE: Normally, fluorescent lights should be turned off only if the "off" period will exceed one hour. Turning them off for shorter periods (thus requiring more starts) could lessen their life and offset any savings. However, when the peak demand, referred to previously, occurs in the middle of the day and "load control" is being practiced, the savings that can be realized through "fly-wheel" effect would more than offset the expense incurred through decreased lamp-life.

Lights — Leaving lights on when not needed, adds appreciably to the cooling load, particularly in those buildings where a high level of illumination has been provided.

Venetian Blinds — Exercise control over their use so that when closed to sunlight they are opened again when the position of the sun has changed and lights are no longer required.

Air-Handlers — Air handlers should be operated only when needed. They should not be operated in non-occupied areas. Pay particular attention to the outside fresh-air dampers, if they are left open when they should not be, they will add to the cooling load, the KW Demand, and KWHs metered.

Nonconditioned Areas — Block off areas that are *not* cooled, from areas that *are* cooled, wherever possible.

Windows and Doors — Keep windows and doors closed.

Insulation — Insulate cold-air ducts, particularly where they pass through, or near, heated or warm areas. The same practice applies to chilled-water lines.

Cold-Air Ducts— Keep clean, and free of obstructions.

Table 9-1. Over-voltage Lamp Relation Table

120-volt lamps	Wattage	Replace by 130-volt lamps**	Wattage	Replace by 125-volt lamps**	Wattage
	10		15		...
	15		*25(e.s.)		...
	25		*40(e.s.)		...
	40		50		...
	50		75		...
	60		75		...
	75		100		...
	100		...		150
	150		...		150
	200		...		200
	300		...		300

* (e.s.) — extended service lamps

** Exceptions:

1. When the existing fixture will not accommodate the larger-wattage lamp in the 130- (or 125-) volt rating, use 130- (or 125-) volt lamp of same wattage as the 120-volt lamp, unless the amount of light available after the substitution would be insufficient for the purpose.
2. When the circuit capacity is reached and the addition of more 130- (or 125-) volt higher-wattage lamps would overload the distribution system, use 130- (or 125-) volt lamps of the same wattage unless light available after the substitution would be insufficient for the purpose.

Openings — When possible, close any openings to basements, elevator shafts, and stairwells leading to floors below, which are not cooled.

Excess Ventilation — Watch for excess ventilation from cooled spaces. Exhaust fans in particular, should be watched.

Economies of Over-voltage Incandescent Light Bulb Use — The use of over-voltage incandescent light bulbs (lamps) in 110V-120V circuits can result in substantial savings over the use of 120-volt bulbs at most facilities. A 130-volt bulb provides 200 per cent longer life and a 125-volt bulb 70 per cent longer life than a 120-volt bulb in the 110V-120V socket.

To achieve an equivalent level of illumination with the 130-volt bulb, it is necessary to change to the next higher wattage bulb — as shown in Table 9-1. Savings are not possible in a conversion to 130-volt lamps over 100 watts because of an excessive consumption of electric energy.

Conversions from 120-volt, 100 watts and over, should be made *ONLY* to 125-volt bulbs as shown in Table 9-1. Levels of illumination will decrease by approximately 12½ per cent in the 150-, 200-, and 300-watt, 125-volt lamps.

Over-voltage Lamps (Man-hour Savings)

The use of 130-volt incandescent lamps in place of the standard 120-volt lamp not only provides a savings in lamp cost due to the longer life of the 130-volt lamps, but also provides an opportunity for substantial savings in man-hours assigned to lamp replacement.

CHAPTER 10

Service Contracts

General

Where trained, qualified personnel are not available or justified, commercial service contracts should be utilized. When soliciting bids, specifications should be incorporated in the contract to assure satisfactory performance. Sample specifications for both refrigeration and water treatment follow.

Specification For Refrigeration Service — This form (See Exhibit 10-A) serves as a standard specification for this type of service and should be made a part of every refrigeration service contract.

Specifications For Water-treatment Service — Exhibit 10-B serves as a standard specification for this type of service. It provides for consulting service by reputable water-treatment companies using chemicals prescribed by them in accordance with their established program.

SPECIFICATIONS FOR REFRIGERATION SERVICE

I. MONTHS COVERED - The months covered under this specification shall be:

____

Quotations shall be submitted covering these months. A quotation for each month is desired, preferably in equal amounts.

II. REFRIGERATION - The following items apply to both Self-Contained Units and Central Systems. (Note that temperature control is covered separately under Item III.)

A. Start of Cooling Season - At the start of the cooling season (see Item I, above) the following services shall be performed by a qualified refrigeration and air-conditioning maintenance contractor:

1. General - Visually and manually check the component parts of the refrigeration cycle, and the system, to determine that all valves and mechanical operators are correctly positioned for the system's mechanical operation.

2. Compressor Oil - Check compressor oil-level.

3. Refrigerant - Check refrigerant level.

4. Cooling Tower - Check cooling tower and fill condensing system. Check if bleed line is tapped into the condensing-water line between the pump and the tower so that water is bled to waste only when the tower is operating. Vacuum clean or wash condenser coils and fins of units using air-cooled condensers.

5. Expansion Tank - Check expansion tank, if provided, for chilled-water level. Where there are direct expansion coils, vacuum clean or wash condenser coils and fins.

6. Operating Check - Place system in operation and for the first 30 minutes, continually check oil-level and refrigerant-level. Add oil and refrigerant ONLY after oil leaks or refrigerant leaks have been found and repaired.

7. Inspection - Perform inspections as called for under Item B.

B. Monthly Inspections - During each of the months set forth in Item I, above, there shall be performed an inspection covering the following:

1. General - Check general performance of refrigerating cycle and cooling system.

Exhibit 10-A. Specifications for refrigeration service.

2. Operating Indicators - Check operating pressures and temperatures of the refrigeration cycle and cooling system.

3. Refrigeration Controls - Check operation of refrigeration controls and reduction devices. Superheat setting of metering devices. (Inlet and outlet of evaporator.)

4. Safety Controls - Check and adjust all safety controls.

5. Refrigerant Leaks - Inspect equipment for refrigerant leaks, using accepted leak-detector. Electronic leak-detector must be used on units exposed to wind. If leaks are noted, provide information on their extent and any repairs necessary.

6. Air Purge - Purge air from system.

7. Lubrication - Lubricate all moving parts as required by manufacturer.

8. Belts - Check and adjust tension and alignment on all belts.

9. Pump Packing - Check pump packing and adjust where necessary.

10. Strainers - Clean all strainers.

11. Valves - Inspect all valves and adjust where necessary.

12. Water Leaks - Check for water leaks.

13. Cooling-Tower Water Level - Check level of water in cooling-tower sump and adjust float valve, if necessary.

14. Bleed - Check condensing-water bleed, usually located at cooling-tower or in the condensing-water line between the pump and the tower. (Do not confuse with tower overflow line.) Check for proper bleed as originally set up. If bleed rate has not been set up properly, note on report the amount of bleed and correction made, together with recommendations.

15. Air Dampers - Check and adjust air dampers for proper air-distribution.

16. Miscellaneous - Make all other necessary adjustments, lubrication, preventive maintenance, etc., required for proper care of the system.

17. Check on Compressor Oil - Make final check on compressor oil and replenish, if necessary.

Exhibit 10-A. (*Continued*) Specifications for refrigeration service.

18. Operating Procedures - Based on the findings revealed by the initial start-up and inspection, review operating procedures and provide necessary written instructions for building maintenance personnel.

C. Inspection Reports - Upon completion of each monthly inspection, the contractor shall prepare a report for the owner (in triplicate) covering the following:

1. General - Show date of inspection and building official or representative contacted. Comment should be made as to the general condition and cleanliness of the entire system.
2. Rusting - Note any rusting and need for touch-up painting.
3. Filters - Note condition of filters.
4. Maintenance Personnel - Do maintenance personnel fully understand the proper operation of the equipment and the routine maintenance that is required?
5. Operating Indicators:
   a. Time of day observed: ____________________
   b. Dry-bulb Temperature: Outside ______ °F; Inside ______ °F
   c. Wet-bulb Temperature: Outside ______ °F; Inside ______ °F
   d. Compressor: Head Press. ______ lbs ______ ozs
Suction Press. ______ lbs ______ ozs
   e. Chilled-Water Temperature: In ______ °F; Out ______ °F
   f. Direct Expansion Systems: Return-air temp. ______ °F
   g. Discharge Temp. at coil ______ °F.
6. Services Performed - List services performed.
7. Instructions - List comments or instructions given to building officials or maintenance personnel and mention the men by name and title. State if previous instructions given have been followed and if the results have been satisfactory.
8. Repairs - List any repair work, replacement of worn-out parts, etc., if required, and approximate cost.

D. End of Cooling Season - At the end of the cooling season, as indicated under Item I, the following shall be performed:

1. Refrigerant Charge - Pump down and store refrigerant charge, if necessary. Close king valve or liquid line shut-off valve.
2. Water - Drain water from all parts of the system likely to freeze.
3. Condenser and Cooling Tower - Inspect, clean and flush condenser and cooling tower.
4. Compressor - Place compressor on pump-down cycle to prevent metal sizing and drying-out of shaft seal.
5. Safety - Prepare a "Red Tag" for each operating component of the system, indicating what was done, and attach it to the control switch or valve of each component.

Exhibit 10-A. *(Continued)* Specifications for refrigeration service.

III. TEMPERATURE CONTROLS

Self-Contained Units - Check and adjust entire temperature-control system in line with the following:

1. Inside Space Thermostat(s) - Adjust to maintain 78 F in all cooled spaces.

2. Outside Air Thermostat(s) - Adjust high limit outside-air thermostat for 75 F setting.

3. Outside Air Damper(s) - Adjust outside-air damper(s) for a minimum setting of 10 to 15 percent outside-air in conjunction with maintaining inside space-temperatures of 78 degrees F.

IV. BID PRICE - The bid shall show the monthly charge for the inspections covered under this bid.

V. EMERGENCY SERVICE - Emergency service shall be available under this contract between inspections. Payment for service caused by improper operation of the system, or breakdowns due to failure to follow written recommendations in a previous report, shall be at current, established hourly-rates.

VI. PARTS REPLACEMENT - If inspection reveals that parts are required, new equipment shall be charged for at published list prices, less applicable discount. Repair or exchange parts shall be at published prices. Wherever possible, parts shall be from the same manufacturer as those replaced. However, such repair or replacement of parts must be authorized by the contracting officer or his authorized representative.

VII. COST BASIS - In addition to the bid price, the bid shall show the hourly rate(s) to be charged for service performed under Items V and VI. The charge for the work performed under these items shall be based only on actual service performed and shall be computed on a time, and fixed hourly-rate, basis. This hourly-rate basis is to be set forth in the bid. Failure to set forth the fixed hourly rate(s) for this service shall be cause for rejection of a bid.

VIII. QUALIFIED PERSONNEL - All inspections and authorized repair work, etc., shall be performed only by qualified personnel.

IX. BUILDING CONDITIONS - Bidders are urged to visit the site to inform themselves as to all existing conditions. Failure to do so will in no way relieve the contractor of any performance required by this specification.

Exhibit 10-A. (*Concluded*) Specifications for refrigeration service.

SPECIFICATIONS FOR WATER TREATMENT FOR AIR CONDITIONING SYSTEMS

1. Months Covered - The months covered under this specification shall be:

   ____

2. Bids - Bids (in quadruplicate) are to be submitted for furnishing water-treatment chemicals and service required in operation of the water cooling tower(s) and chilled-water circulating system(s). Treatment, chemicals, and service shall include periodic analyses of the condensing and chilled water; furnishing chemicals needed to inhibit scale and corrosion and control slime and algae in a manner to avoid deterioration of the systems; prescribing the proper dosage of chemicals; and instructing the operating personnel in their use.

3. Qualifications - Proposals will be considered only from established firms which are currently engaged in the field of industrial water chemistry and have suitable laboratory facilities for water analyses and evaluation. Service to be performed shall include inspection of the air conditioning plant by a qualified chemist or engineer, analyses of water samples, written reports—including analytical results, instructions to personnel, and appropriate recommendations.

4. Visit to Site - Bidders are urged to visit the site to inform themselves as to all existing conditions. Failure to do so will in no way relieve the contractor of any performance required by these specifications.

5. Initial Water Analysis - Bidder shall submit with his bid an analysis of the local water and a statement of the chemical compounds he proposes to use in performance of the contract.

6. Award Conditions - Upon award of contract, contractor shall make an initial visit to the premises for the purpose of instructing operating personnel and thereafter he shall make at least one inspection of cooling tower, condensing- and chilled-water systems each month during the cooling season, to check operation, instruct personnel, and collect water samples for analysis. Upon completion of monthly inspection, contractor shall furnish the contracting officer a report (in triplicate) on the general condition of the system, the instructions issued to operating personnel, and the analysis of the water samples together with recommendations.

7. Monthly Reports - The reports shall include all pertinent items which may be needed to properly evaluate the water-treatment program.

8. Feed or Test Equipment - Contractor shall furnish without cost any chemical or test equipment not already at the building, which may be needed to carry out the water-treatment program.

Exhibit 10-B. Specifications for water treatment for air-conditioning systems.

CHAPTER 11

Safety

General

The matter of safety cannot be stressed too strongly. While the items covered herein are mainly geared to air conditioning, they also include some that are general in nature yet deemed all-important.

Accidents — Accidents resulting in personal injury can almost always be attributed to unsafe work practices by either operating or maintenance personnel. Accidents often happen because personnel concerned do not realize the danger of improper practices. Frequently, the proper practice is known, yet ignored because the employee is in a hurry and is careless.

Unsafe Conditions — Maintenance personnel can contribute most toward the success of safe operation since they know the dangers resulting from incorrect use and maintenance. They should be trained to recognize and promptly report any unsafe condition arising from deficiencies in the equipment or poor operating practice.

Exception — Where a rule cannot be complied with literally because of practical difficulty, or where its liberal application would cause undue hardship, the enforcing authority may, upon proper application, grant exceptions but only when it is clearly evident that reasonable safety is assured.

Common Housekeeping Safety Rules

Storing Material — To avoid accidents from falling objects, store material in an orderly fashion. Insure that pipes, ladders, barrel racks, etc., do not extend into aisles. Use blocks or racks to prevent pipes from rolling. Dispose of scrap, and store waste, lubricants, rags, and other flammables in proper containers provided for the purpose.

On the Job — Maintain good housekeeping during progress of any work. Keep oil cans or wiping cloths away from moving machinery. Upon completion of a job, all surplus material shall be removed and proper disposition shall be made of these.

Clothing — Use appropriate wearing apparel or special protective clothing or safety equipment when required. Loose clothing, neckties, unbuttoned long sleeves, etc., should be avoided.

Tools — Keep tools in good condition and use them only for the purpose for which they were made. Dull, worn, or defective tools should be removed from use and repaired or replaced before being reissued. Maintenance personnel should have a thorough knowledge of how to use power tools properly. Pick up tools when finished and return each to its proper place.

Lifting and Pulling — Hand trucks, chain falls, and other types of lifting or conveying equipment should be used whenever practicable for handling large pipe, barrels, heavy equipment, etc. Safe individual and team effort should be promoted.

Guards or Barriers — Provide physical barriers or otherwise guard all hazardous conditions during the progress of a job until normal conditions are restored. Replace permanent guards or other safety devices which have been removed from the equipment. Do this before leaving the job.

Safety Devices, Positions, and Equipment

Safety Devices — Examine safety devices before they are used to make sure they are in good condition. Make certain the proper device is used for the particular work to be done.

Safe Position — Place yourself in a safe position while working to avoid falling, stumbling, slipping, or moving backwards against live parts. There is no reason to take chances that will endanger yourself or others.

Protective Equipment — Personnel protective equipment such as: goggles, masks, gloves, aprons, etc., shall be used when prescribed and supervisors shall arrange to store them properly and issue them as needed.

Ladders and Scaffolding — Ladders should never be spliced together. Makeshift extensions are dangerous. Broken, cracked, or defective ladders should be taken out of use and repaired immediately. Keep ladders clean and free from dirt or paint which may conceal defects. Ladders and scaffolds shall be of a type and size suited to the job as prescribed by the supervisor. All ladders shall have nonslip feet. Extension ladders shall have at least a 3-foot lap. Secure the top of straight and extension ladders where there is danger of the ladders slipping. Place the bottom of such ladders not more than ¼ of their height from the perpendicular. Both hands shall be free for climbing. Tools shall be hoisted with a handrope or carried on a tool belt. Rolling scaffolds, wheels, and swivels shall be locked when scaffold is in use, and shall not be moved with employees or tools and equipment on the platform. Planks used in temporary scaffolds shall be straight-grained, free of defects, not less than 2 inches in thickness, and supported at not more than 8-foot centers.

Electricity Safety Measures

Precaution Against Live Circuits — Avoid use of steel rules, brass-bound rules or metal measuring tapes while working on or near live parts. Assume that every circuit is "alive" until checked with a voltmeter, a short-circuiting device, or other authorized means.

Grounding — Ground all hand electric-power tools before use.

Fuses

a. *Removing* — Before removing fuses, open switches fully. To remove a fuse from a circuit carrying field-current without first opening the switch, is particularly dangerous.
b. *Circuit under 500 Volts* — Use an approved low voltage fuse-puller to remove fuses on a circuit of less than 500 volts where no switch is provided.
c. *Circuit with 500 Volts or More* — Use an approved, insulated high-voltage fuse-puller to remove fuses on circuits of 500 volts or more where no switches are provided.
d. *Breaking Contact* — Remove fuses by first breaking contact with the hot side of the circuit. Use the reverse procedure when replacing fuses. Insert the fuses in the cold terminal first.

One Hand Rule — Rigidly observe the "one-hand" habit (placing the other hand out-of-use in a belt or pocket) when throwing open-type switches, removing leads from apparatus or a terminal board, pulling plug leads from a distributing board, operating line power rheostats, and other similar operations. Never grasp 2 sides of a circuit, alive or dead, at the same time.

Applicable Safety Devices

The safe practices presented here are those which should be commonly and habitually executed by employees engaged in heating, ventilating, and air-conditioning work.

Acids, Solvents, and Other Chemicals — Avoid spilling and splashing and use protective equipment specified by the supervisor. Wash hands after using. If clothing becomes saturated with acids or chemicals that may cause injury, wash affected parts of the body with soap and water, change clothes, and report to the supervisor.

Caustic Soda (Lye) — Caustic soda (lye) must be handled with care. When added to water a vigorous reaction takes place. If done carelessly and the solution splatters on the hands or face, a serious burn may result. For protection, use gloves which can be obtained from any local market. Use chemical type goggles, obtained through local purchase, if deemed necessary. Using a pail or bucket with water in it, add the caustic gradually and allow the solution to cool to permit easy handling of the container. The same precaution must be utilized in the use of chemicals containing acids such as sulfuric, chromic, etc.

Ventilation — Proper ventilation must be provided in all cases.

Shut-Off Precautions — Shut off and tag all valves or controls regulating the flow of steam, condensate, refrigerants, gas, etc., before disconnecting piping or working on equipment. Steam, gas, and refrigerant lines shall be carefully examined before proceeding with the work; to insure that valves are holding, that pressure is absent, and that the piping is properly drained or vented. Tags shall be removed and service restored only by the person who originally tagged the valves or at his request.

Soldering, Brazing, Cutting, and Welding — Use personal protective equipment prescribed for the specific job. No flame cutting and welding shall be done outside shop space without the knowledge and consent of the supervisor. Barriers shall be provided to confine sparks and injurious light and to restrict occupants or the public from hazardous areas. The area shall be checked before operations are started and suitable precautions taken to prevent fire. A fire extinguisher of the proper type should be readily available. The area shall be thoroughly checked for signs of fire after the work is completed. Store and transport compressed-gas cylinders so they cannot fall or be knocked over. Keep protective caps on cylinders when they are not in use

Threaded Pipe — Use caution when cutting or threading pipe and be alert to the danger of cuts. Cap or wrap the ends of threaded pipe while transporting or handling it.

Cleaning with Compressed Air or Steam — Dusting with compressed air or cleaning with steam shall be performed only when authorized by the supervisor. Compressed air shall not be used for dusting clothing.

Testing for Leaks — Protect hands and face when testing for leaks in steam, gas, or refrigerant piping and use approved leak detectors.

Refrigerant Safety Rules

Nameplate Check — It is always good practice to check the compressor name-plate data before charging or adding refrigerant to make certain of the correct *amount* of refrigerant in the system and the *kind* of refrigerant.

Opening Refrigerant Drums — Never open drums of refrigerant to the atmosphere when it is above 74° F. At temperatures above this, the drum will be under pressure and could result in personal injury.

Torches — Never use a torch or attempt to repair a line containing refrigerant until it is certain that all gas has been pumped out of the section of pipe to be repaired and the line has been valved off.

Refrigerant Cylinders — To provide for some margin of safety:

a. Never heat a cylinder above 125° F.
b. Never apply a direct flame to a cylinder.
c. Never place an electric resistance heater in direct contact with a cylinder.
d. Never transfer refrigerant from one container to another.
e. Avoid dropping cylinders. They should never be rolled on their sides. Valves may be damaged and the cylinders dented.
f. Always replace valve cap and hood cap when cylinder is not in use or when cylinder is empty.
g. Secure all cylinders in upright position with chain or strap when they are not mounted in suitably-constructed stand for withdrawal or for refrigerant service.

Fresh-Air Intakes and Ducts Safety Measures

Fresh-Air Intakes — Do not permit accumulation of combustible material near the intake, particularly when near buildings or structures where there is a possibility of smoke or fire being drawn in.

Ducts — Large accumulations of dust and waste in ducts should not be permitted.

Repair Equipment — Equipment used in making repairs should be equal or better than the equipment replaced. "Makeshift" repairs are not to be made. There may be occasions where a temporary repair is necessary due to circumstances. In these cases, safety is not to be compromised.

CHAPTER 12

A Program for Controlled Operation of Large Heating and Cooling Plants

Controlled Operation

Controlled operation means intelligent operation. The fact that modern plants are equipped with automatic features does not mean that proper or economical operation is automatically assured. It is too often the case that controls were installed improperly and do not meet the original design conditions or do not meet actual conditions required by the building(s). Thus, some means must be provided to assure power plant operation. The program described below, has been designed to specifically perform this function. It will supply much of the data available only through costly control centers and at a fraction of their charges. At those installations where such centers already have been installed, it will provide important supporting data probably not available from the center. Thus, the program will assure efficient plant operation and result in worthwhile dollar savings. To accomplish this, two special forms have been developed: A Heating-Cooling Chart (Exhibit 12-A); and a Refrigeration Plant Operating Log (Exhibit 12-B).

Heating-Cooling Chart

Purpose — The purpose of this chart is to provide building superintendents with a graphic representation both of their heating and cooling plants. This is accomplished by plotting the actual daily consumption of fuel during the heating season and "ton-hours" during the cooling season against established targets for corresponding mean temperatures. The chart monitors plant operation in that the operator is made aware of excess energy used whether in fuel or electricity. Thus warned, it prompts investigation for cause, which in turn, uncovers deficiencies and malfunctions in both operation and equipment that would not be noted otherwise. Its use during the cooling season in conjunction with the Refrigeration Plant Operating Log provides a relatively simple, easily understood means for management to achieve operating efficiency and savings, otherwise obtainable only through the use of costly and complex electronic control centers.

Plotting — It is important that plotting be done daily. In this way any unusual conditions (whether external or internal) that could have affected building operation, may be easily recalled. (See "D," in Exhibit 12-A.)

Interpretation — For best results, the chart should show some red. Upon investigation for cause, and resultant corrective action, the purpose of the chart is accomplished.

Heating and Cooling Targets

New Buildings — For new buildings or older buildings where good fuel consumption records are not available, the base "fuel consumption target" may be obtained by using the Fuel Consumption Chart. (See Exhibit 12-C.) An example illustrating the use of this chart is included on the chart. The base fuel-consumption target is the figure inserted in column (1) opposite the mean temperature of 64° F, shown in column (2) of the Heating-Cooling Chart (Exhibit 12-A).

Other Buildings — For buildings where good fuel-consumption records have been maintained, the base target for a start can be obtained by dividing the total fuel consumption for the past year by the total degree days in the locality for that year, and then multiplying the result by the normal degree days. The result will be the target for insertion in column (1) opposite the 64° F mean temperature shown in column (2). An example of how column (1) is completed is shown in the lower half of the reverse side of the Heating-Cooling Chart.

Target Adjustment — See item "F" on the reverse side of the Heating-Cooling Chart.

Refrigeration Plant Operating Log

Purpose — The purpose of this form (Exhibit 12-B), is to provide pertinent operating data to a supervisor or operator of a central air conditioning system, for evaluating the daily performance of his refrigeration plant. Through proper evaluation of the data and resultant action dictated by current and anticipated local climatic conditions, the log form will insure the most efficient and economical plant operation. The electrically-driven plant is used as an example since it represents the type most widely used at present. Comparable savings are possible with other type plants through similar control over operation, since "energy" is required whether it be electricity, gas, oil, or steam.

Example — Assume a refrigeration plant consists of two 1,000-ton centrifugal compressors. Under usual design conditions, the No. 1 or lead machine takes care of approximately 65 per cent of the cooling requirements. This means that, ordinarily, the No. 2 or lag machine would be cut-in at this point and thus both machines could be on the line during a period when the demand for electricity by *other* than the air conditioning equipment was at its maximum. Thus, the total demand for electricity could be increased by as much as 1,000 KW when the No. 2 machine cut in. Depending upon the Demand Charge rate (which could be from $1.00 to $1.50 per KW), the bill for the month would be increased $1,000 to $1,500 for that month even though the No. 2 machine operated only 15 minutes (in the case of a 15-minute demand rate), and then only for that one brief time during the entire month while bringing the discharge chilled-water temperature down to the cut-out point. In some cases, the rate structure calls for billing the demand charge at this "peak" demand for the subsequent 11 months even though lower demands are actually registered.

Fly-Wheel Effect — With "load control" in mind, the "key" is in making the chilled-water in the piping and coils throughout the building act as a "fly-wheel" in carrying the building over a period of peak electric demand in lieu of operating one or more compressors during this period. Success in accomplishing this depends mainly on the amount of water in the chilled-water system, the Wet-Bulb temperature at the time, the duration of the peak demand period or conditions peculiar to a particular building. (This action is covered thoroughly in Chapter 9.)

a. *Chilled-Water Temperature* — Assume the chilled-water pumps are being operated in accordance with design instructions. To assist in obtaining the "fly-wheel" effect, the chilled-water supply temperature-control point could be set 2 to 3 degrees lower than the actual design-control point, during periods of low load for one machine. Do not change the freeze cut-out point and watch the operation carefully to make certain that this action does not activate the freeze cut-out control.

b. *Effect* — This sub-cooling of the chilled water for circulation would have the effect of lengthening the period required for the chilled-water return temperature to rise to the point where it would become necessary to use two machines.

Plant Capacity — Assuming the plant is functioning properly to the best of the operating personnel's knowledge, it will be necessary to operate the plant for several weeks to ascertain the capabilities of the plant. In other words, it is necessary for the operator to know how long he can utilize the "fly-wheel" effect, and under what conditions of Dry- and Wet-Bulb temperatures. With this knowledge and assuming he knows when his peak KW demand occurs during the day, he can schedule the operation of his refrigeration plant accordingly.

HEATING AND COOLING CHART
(See instructions on reverse.)

CITY | STATE | MONTH AND YEAR

BUILDING | MEAN TEMPERATURE | DEGREE DAYS

TARGETS (24 hours)				DAILY RECORD																														
HEATING		COOLING																																
	MEAN TEMP.	TON HOURS	MEAN TEMP.																															
(1)	(2)	(3)	(4)	1	2	3	4	5	6	7	8	9	10	11	12	13	14	15	16	17	18	19	20	21	22	23	24	25	26	27	28	29	30	31
	-18																																	
	-17																																	
	-16																																	
	-15																																	
	-14																																	
	-13																																	
	-12																																	
	-11																																	
	-10																																	
	-9																																	
	-8																																	
	-7																																	
	-6																																	
	-5																																	
	-4																																	
	-3																																	
	-2																																	
	-1																																	
	0																																	
	1																																	
	2																																	
	3																																	
	4																																	
	5		110																															
	6		109																															
	7		108																															
	8		107																															
	9		106																															
	10		105																															
	11		104																															
	12		103																															
	48		67																															
	49		66																															
	50		65																															
	51		64																															
	52		63																															
	53		62																															
	54		61																															
	55		60																															
	56		59																															
	57		58																															
	58		57																															
	59		56																															
	60		55																															
	61		54																															
	62		53																															
	63		52																															
	64		51																															
	65		50																															
DATE →				1	2	3	4	5	6	7	8	9	10	11	12	13	14	15	16	17	18	19	20	21	22	23	24	25	26	27	28	29	30	31

COMMENTS

EMPLOYEE IN CHARGE OF HEATING PLANT

TITLE

Exhibit 12-A. Heating and Cooling chart.

SAMPLE CHART

Assume a building using oil has a Consumption Factor of 15 Gals Oil Per Degree Day and the actual consumptions for the first 5 days of a heating month are as listed below. The figures in column (1) on the sample chart would be as shown and the plotting of the chart for these 5 days would be as illustrated. This follows the explanation set forth under Heating (E) (1) and (2).

NOTE: Column (3) is shown blank on the sample chart since the chart is being used as a heating chart. When being used as a cooling chart, column (1) would be blank and column (3) would have figures in it beginning with a mean temperature of 61° as explained under (B)(2). The actual plotting of the chart is the same whether it be for heating or cooling. The only difference is that during the heating season columns (1) and (2) are used and during cooling season columns (3) and (4) are used.

DATE	MEAN TEMPERATURE	FUEL CONSUMPTION TARGET	FUEL CONSUMPTION ACTUAL
1	25	600	612
2	27	570	553
3	33	480	528
4	30	525	499
5	21	660	647

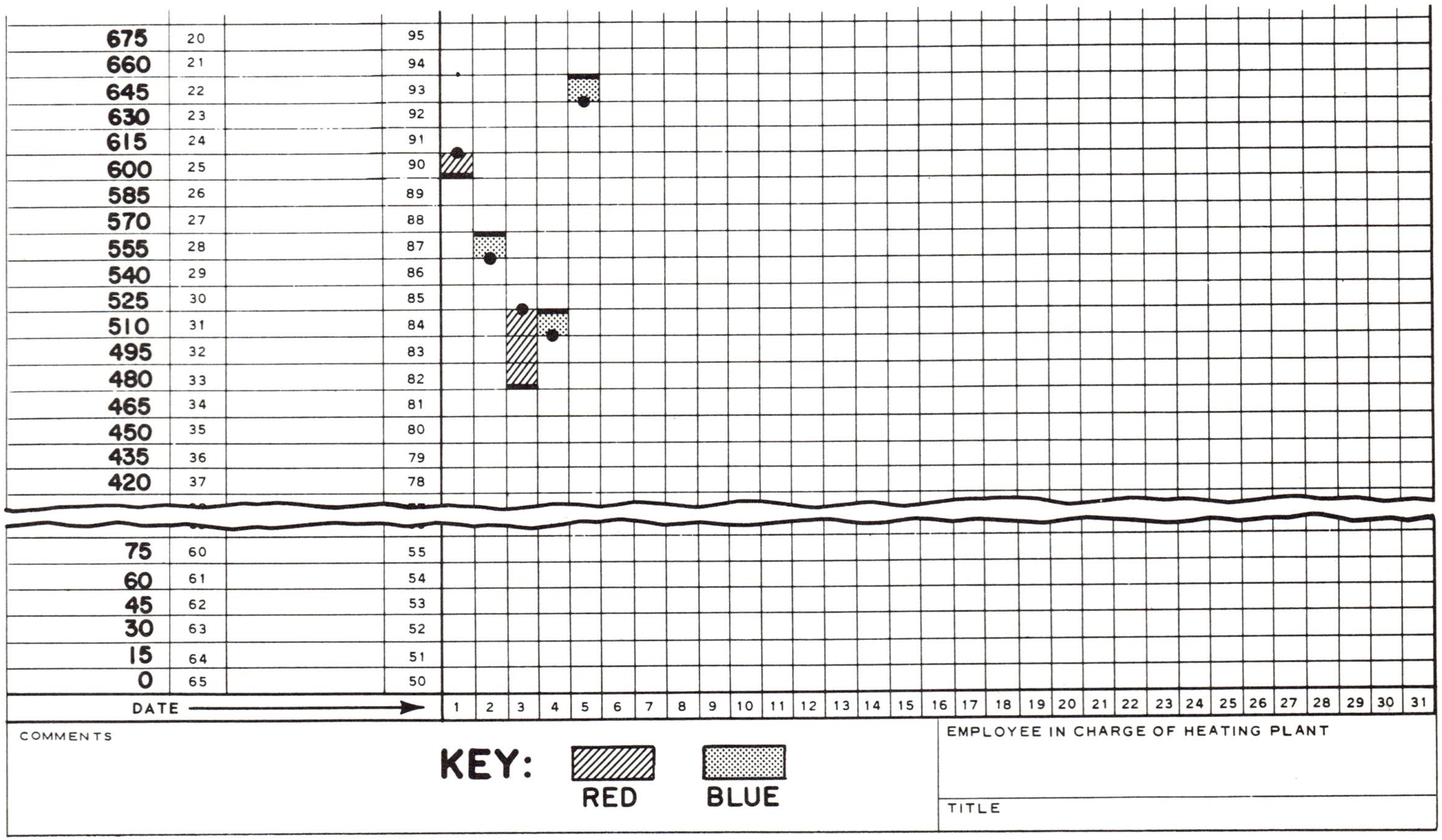

INSTRUCTIONS

A. PURPOSE - The purpose of this chart is to provide the operator of a building with a graphic presentation of both his heating and cooling plants. This is accomplished by plotting the actual consumption of fuel (during the heating season) and the refrigeration ton-hours (during the cooling season) against established targets based on mean daily temperatures.

B. TARGETS - The original targets are set up based on past records of through use of Exhibit 12-C for heating; or Exhibits 12-D and 12-E for cooling. The targets established will be used pending demonstration through actual use that they are not realistic due to local conditions peculiar to a particular building.

1. Heating - The targets for heating are set up in column (1) and the mean, outside, dry-bulb temperatures (hereafter referred to as mean temperature) to use with the heating targets are shown in column (2).

NOTE: The first heating target is set up opposite a mean temperature of 64°. This is based on the assumption that, normally there is sufficient heat generated within large buildings to maintain satisfactory inside temperatures without using the heating plant until the mean outside temperature goes below 65°.

2. Cooling - The targets for cooling are set up in column (3) and the mean temperatures to use with the cooling targets are shown in column (4).

NOTE: The first cooling target is set up opposite a mean temperature of 61°. This is based on the assumption that the internal heat load in large post offices is such that cooling is normally required when the mean temperature exceeds 60°.

C. WHAT IT DOES - The chart monitors building plant operation in that the engineer is made aware of excess energy use whether it be fuel or ton-hours. Thus warned, it prompts investigation of deficiencies and malfunctions in both operation and equipment that would not be noted otherwise. It provides a check in revealing sub-par performance requiring corrective action. Its use during the cooling season in conjunction with Exhibit 12-B, Refrigeration Plant Operating Log, provides a relatively simple, easily understood means of providing management by exception in achieving operating efficiency and savings obtainable otherwise only through the use of costly and complex electronic control centers.

D. DAILY PLOTTING - (Begin with first day of heating or cooling)–

1. Mean Temperature - Ascertain the mean temperature for the day (24 hrs). Place a horizontal line thus (–) for that day of the month under the heading "Daily Record," opposite the same mean temperature shown in column (2) if during the heating season and column (4) if during the cooling season.

NOTE: Where possible, use the mean temperature as reported by the U.S. Weather Bureau. If the temperature cannot be obtained from this source, or from the official cooperative observer for the Weather Bureau for a particular locality, use the mean temperatures as reported in a local newspaper; or by taking the average temperature for 24 hours; or by averaging the maximum and minimum temperatures for the 24-hour period. If this source is used, the same newspaper and the same edition must be used so that the mean temperatures so obtained will cover the same 24-hour period each day.

2. Energy (Fuel Consumption and Ton-Hours)

a. Heating Season - Ascertain the fuel consumption for the day (24 hrs) from boiler room records. Place a small circle or dot thus (o) for that day of the month under the heading "Daily Record," opposite the target figure shown in column (1) nearest in amount to that of the actual consumption. If the circle or dot used to represent the actual consumption is above the mean temperature line for the day, shade the space red between the line and the dot; if below, shade the space between in blue. (See plotting of sample chart at end of these instructions.) If for any day, after the beginning of the heating season, the heating plant is not operated, the words "no heat used" should be inserted in the column for the day. In the case of cooling, the notation would be "no cooling used."

NOTE: For plotting, use only that fuel consumption used in the heating plant. Fuel used for heating water for domestic purposes is not to be included. Where all or part of the domestic water needs is obtained through the operation of the main heating boilers during the heating season, an amount representing the average use for hot water purposes should be ascertained and deducted from the daily consumption and the net amount used to plot the chart.

3. Cooling - Ascertain the total ton-hours for the day from the Refrigeration Plant Operating Log. Proceed to plot in same manner explained under D(2) utilizing red and blue coloring except that column (3) ton-hours is used as a target whereas for Heating, column (1) is used. (See sample chart at end of these instructions.)

E. INTERPRETATION - It is not expected the actual fuel consumption during the heating season or the actual ton-hours during the cooling season will correspond exactly with the target set up for each mean temperature, due to various outside and inside conditions that affect the heating and cooling of a building from day to day. However, when a day shows 5 or more blocks of red, or extended periods of red occur, investigation should be made immediately for cause, and prompt corrective action taken. The optimum chart would be one with the smallest possible variations from the targets and thus have very little red or blue showing. Use space at bottom of chart headed "Comments" to explain excess blocks of red.

F. TARGET ADJUSTMENT

1. Red - When a chart shows a preponderance of red, despite all efforts to determine and correct any operating deficiencies or malfunctioning equipment, the factor should be increased to where the chart starts to show some blue. In this way the days of real excess consumption will stand out and, with repeated concentration and effort, the real cause may be ascertained.

2. Blue - Similarly, when a chart shows a proponderance of blue, the consumption factor should be lowered to where the chart starts to show some red.

3. Purpose - It must be remembered the targets serve their purpose best when the chart shows some red and some blue, so that the excesses will stand out whether they be fuel consumption or ton-hours.

Exhibit 12-A. (*Continued*) Heating and Cooling chart, reverse side.

CITY | STATE | DATE

BUILDING

REFRIGERATION PLANT OPERATING LOG

(See Instructions on reverse)

MACHINE ROOM ACTION						OUTSIDE CONDITIONS			REFRIGERATION PLANT										COMMENTS
									COMPRESSOR		CHILLED WATER				REFRIGERATION LOAD				TOUR #1
											WATER TEMP.								
MACH NO.	PLACED ON LINE	TAKEN OFF LINE	ROUTINE CHECK	TIME OF ACTION	HOURS MACH. ON LINE	DRY BULB *(D.B.)*	WET BULB *(W.B.)*	% SUN-SHINE	MOTOR AMPS *(If electric drive)*	HEAD PRESS	IN	OUT	DIFF.	WATER FLOW G.P.M.	TONS OF REFRIG.	% MACH. CAP.	% INST. REFRIG.	TON-HOURS	
1	2	3	4	5	6	7	8	9	10	11	12	13	14	15	16	17	18	19	
AV. D-B- & W-B- TEMPS.															TOTAL TON-HOURS				

COMMENTS:

TOUR #1

CHECK IF CORRECTIVE ACTION REQUIRED ☐

TOUR #2

CHECK IF CORRECTIVE ACTION REQUIRED ☐

TOUR #3

CHECK IF CORRECTIVE ACTION REQUIRED ☐

ACTION TAKEN BY PLANT SUPERVISOR TO CORRECT CONDITION(S) NOTED ABOVE:

TOUR 1	TOUR 2	TOUR 3	EVALUATED BY:
FROM: TO:	FROM TO:	FROM: TO:	NAME
PEAK DEMAND ______ KW	PEAK DEMAND ______ KW	PEAK DEMAND ______ KW	
TIME OCCURRED ______ AM ______ PM	TIME OCCURRED ______ AM ______ PM	TIME OCCURRED ______ AM ______ PM	JOB TITLE
NAME / TITLE	NAME / TITLE	NAME / TITLE	

Exhibit 12-B. Refrigeration plant Operating-log form.

INSTRUCTIONS

1. The purpose of this form is to provide the operator with pertinent operating data for evaluating the daily performance of his refrigeration plant. The correct interpretation of the data and resultant proper action dictated by current or anticipated local conditions, will provide both efficient and economical operation. Important - During the tour, any system or equipment malfunction noted which requires corrective action beyond the capability or authority of the tour supervisor, shall be so indicated under comments and the block checked (×). The plant supervisor shall evaluate the report and list the action taken to correct the condition(s).
2. Column 1 *(Machine Number)* - Identifies the machine(s) to which the readings pertain.
3. Columns 2, 3 and 4 *(Machine Action)* - Indicate the fact a machine has been "placed on the line" or "taken off the line" or a "routine check" has been made. Use check (✓) to indicate. NOTE—"Routine Checks" shall normally be made at approximately 4-hour intervals.
4. Column 5 *(Time of Action)* - Show time of day "action" indicated took place.
5. Column 6 *(Machine Hours)* - Show number of hours each machine is on the line.
6. Columns 7 & 8 *(Dry-Bulb and Wet-Bulb Temps.)* - Show Dry-Bulb and Wet-Bulb Temps. at time of action, using building instruments, if provided. If you have no instruments, information is usually available from local Weather Bureau Office through automatic telephone service in the larger cities, or is available from the newspaper. If Relative Humidity is reported by Weather Bureau in lieu of Wet-Bulb Temp., obtain W-B Temp. from Psychrometric Chart for the D-B Temp. and Relative Humidity reported. The average D-B and W-B Temps. for the day may be averaged from the figures in the columns *(or the average reported by the Weather Bureau may be used)* and inserted in the spaces at the bottom of the columns.
7. Column 9 *(Percent Sunshine)* - Show percent of sunshine at time of action. This may be estimated at 0, 25%, 50%, 75% or 100%.
8. Column 10 *(Motor Amps)* - Show running amps for the compressor(s) in use.
9. Columns 11–14 - The information to be shown in these columns is taken directly from the indicators provided for the equipment.
10. Column 15 *(G.P.M.)* - Where a Water Flow Meter has been provided for the chilled water, show the G.P.M. If no meter has been provided, obtain the G.P.M. from the performance curve for the pump(s) involved. If 2 or more pumps are in use at the same time and only one refrigeration machine, add the G.P.M. of each and show total G.P.M. only.
11. Column 16 *(Tons of Refrigeration)*

    Based on G.P.M. - If no Btu Meter has been provided, obtain tons from G.P.M. through use of this formula: $\frac{\text{G.P.M.} \times \text{Temp. Diff.}}{24}$

**12. Column 17 *(Percent of Machine Capacity)* - Obtain by dividing tons of refrigeration in Column 16 by rating of refrigeration machine in tons of refrigeration.

**13. Column 18 *(Percent of Installed Refrigeration Capacity)* - Obtain percent of total load by dividing Tons of Refrigeration *(Col. 16)* by Total Tons of Refrigeration Installed.

14. Column 19 *(Ton-Hours)* - Multiply the hours each machine is on the line *(Column 6)* by the corresponding tons of refrigeration *(Column 16)* and insert the product in column 19. At the end of the day, total column 19 and insert the total in the space provided at the bottom.
15. Tour Information:
    a. Tour Period - Show covered by each tour.
    b. Peak Demand - The peak KW demand is obtained from the demand chart *(when one is provided)* at the end of each tour and inserted in the space provided, including the time of day the peak occurred.
    c. Name and Title - Insert name and title of man responsible for the operation of the equipment during each tour.

**NOTE: Complete Columns 17 and 18 only when Ton-Hours for a particular day exceed the "target" by more than 15 percent.

METER READINGS *(See Note Below)*

1. GENERAL - Blocks are provided below for recording meter readings for the various types of "Energy" used, including a block for obtaining gallons of fuel oil used. Meter readings are taken at end of Tour 3 only. NOTE - A separate block is provided for obtaining the consumption for refrigeration service when this service is sub-metered.
2. METER CONSTANT - Normally meters have a "constant" or "multiplier." The difference in meter readings *(Present reading minus previous reading)* is multiplied by this constant to obtain the actual energy consumption. Make certain that this constant is applied.
3. MULTIPLE METERS - Where more than one meter is used to measure the same energy, add the consumptions to obtain the total amount of that energy used and insert this total in the space provided below the appropriate block. EXCEPTION - Do not add the consumption of a sub-metered energy.
4. NOTE: Complete applicable blocks only where installation utilizes "Total Energy" concept, such as all electric, all gas, or on-site electric generation.

METER READINGS AND ENERGY CONSUMPTIONS

* UNIT	* UNIT	* UNIT
METER NO. ______	METER NO. ______	METER NO. ______
READ. END PRES. TOUR 3 ______	READ. END PRES. TOUR 3 ______	READ. END PRES. TOUR 3 ______
READ. END PREV. TOUR 3 ______	READ. END PREV. TOUR 3 ______	READ. END PREV. TOUR 3 ______
DIFFERENCE ______	DIFFERENCE ______	DIFFERENCE ______
DIFF. × ______ CONST. = ______	DIFF. × ______ CONST. = ______	DIFF. × ______ CONST. = ______

* UNIT	* UNIT	* UNIT
METER NO. ______	METER NO. ______	METER NO. ______
READ. END PRES. TOUR 3 ______	READ. END PRES. TOUR 3 ______	READ. END PRES. TOUR 3 ______
READ. END PREV. TOUR 3 ______	READ. END PREV. TOUR 3 ______	READ. END PREV. TOUR 3 ______
DIFFERENCE ______	DIFFERENCE ______	DIFFERENCE ______
DIFF. × ______ CONST. = ______	DIFF. × ______ CONST. = ______	DIFF. × ______ CONST. = ______

*Indicate if Cu. Ft. (Gas) of lbs. (steam)

SEPARATE REFRIG. SERVICE

* UNIT
METER NO. ______
READ. END PRES. TOUR 3 ______
READ. END PREV. TOUR 3 ______
DIFFERENCE ______
DIFF. × ______ CONST. = ______

*Indicate if KWH, Cu. Ft. Gas, lbs. (steam)

FUEL OIL

BASED ON GAL'S. IN TANK
AT END OF PRES. TOUR 3 ______
AT END OF PREV. TOUR 3 ______
GALS USED = ______

TOTAL HEATING ______

Exhibit 12-B. *(Continued)* Refrigeration plant Operating-log form, reverse side.

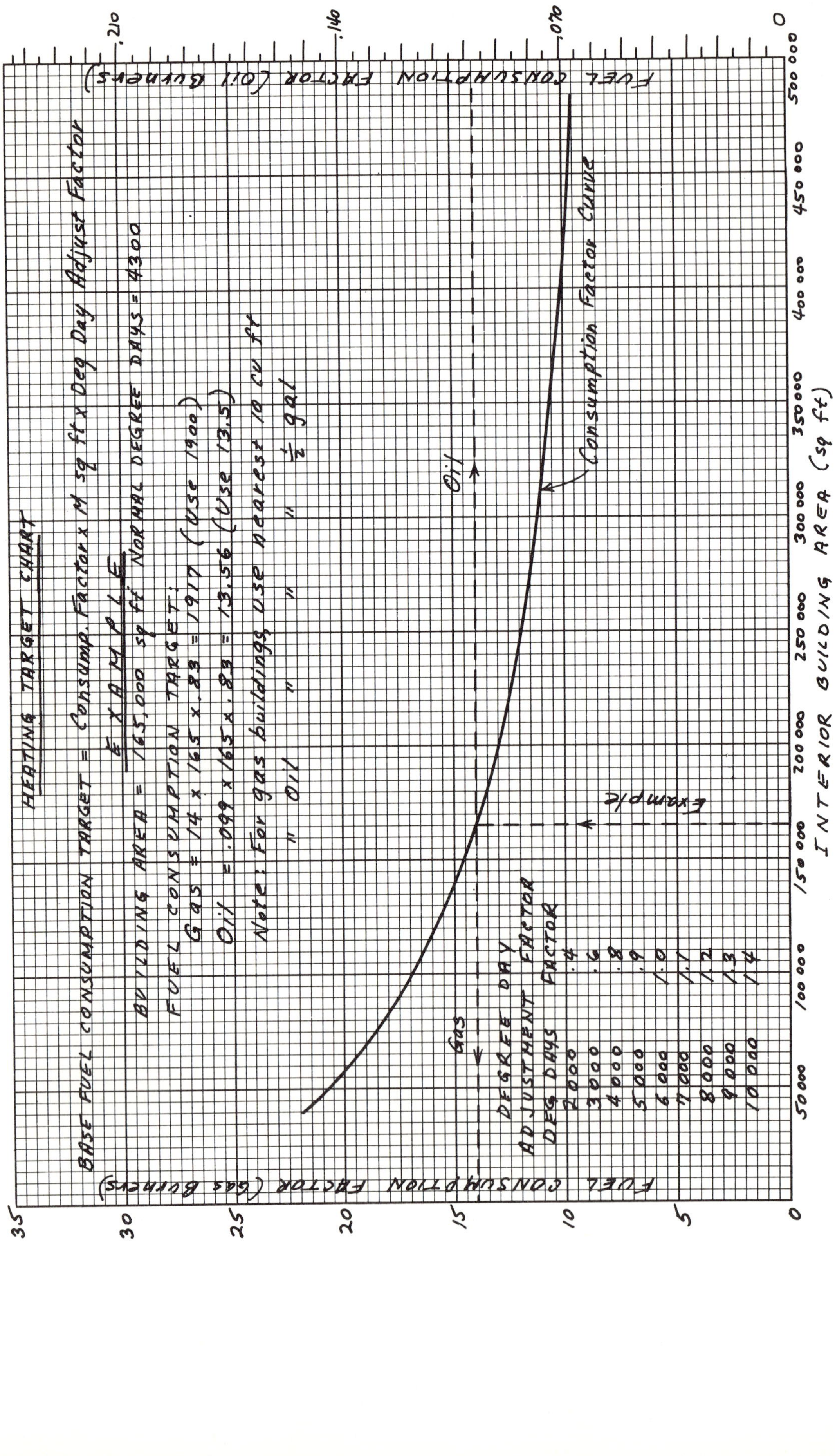

Exhibit 12-C. Heating target chart for interior building area.

Operating Indicators (See Exhibit 12-B.)

Column 10 (Motor Amps) — The entries in this column show the *running* amps of compressor motor(s) in use. The full load, name-plate ampere rating of the motor(s) together with their overload trip rating should be noted in order to know when and for how long one or more compressors can be utliized before having to cut-in additional refrigeration capacity. This knowledge is necessary whether the lag machine is controlled manually or automatically.

Column 12 (Return Chilled-Water Temperature) — As stated previously, the return water temperature of the chilled-water system is the key to the successful use of the log form. From observation, as warm weather is experienced during the first several weeks of operation, the operator will soon learn to what extent he can employ the "fly-wheel" effect. Normally this should be in direct relation to the Dry-Bulb temperature except where conditions local to the area (such as low Wet-Bulb depression resulting in high humidity, or to a particular system, exist, that would have an undue influence. By the time extended hot weather is encountered, the operator should be experienced in interpreting the temperatures in Column 12 against the running amps in Column 10, to the point where he would know when he needed 1 machine and 1 pump, 1 machine and 2 pumps, or both machines and both pumps to carry the load over the desired period where plant equipment permitted this degree of flexibility.

Column 17 (Per cent of Machine Capacity) — This column is a check on proper operation and the resultant efficiency of the refrigeration plant. For example, if the running amps as shown in column 10 are averaging 80 per cent of the full load amps, and the corresponding figures in column 17 are about 50 to 60 per cent, something is wrong.

Column 19 (Ton-Hours) —Ton-hours are used as a target. It represents the time a refrigeration machine is on the line and the tons of refrigeration it was producing during that time(s). It is tied-in directly to the differential between the temperature of the water leaving and returning to the chiller. The method of establishing the targets is covered later, under the heading "Cooling Targets."

Interpretation — The actual ton-hours are compared daily against the target ton-hours inserted previously in column (3) of the Heating-Cooling Chart. When the actual ton-hours exceed the target for a particular mean temperature by 20 per cent or more, action should be taken to ascertain the cause. Several possible causes for Target ton-hour discrepancies are listed below:

a. *Humidity* — It will be noted that ton-hours are tied-in with Dry-Bulb temperatures. For some days, excess ton-hours may be attributed to high humidity. This may show up as the reason when the difference between the Dry- and Wet-Bulb temperatures (referred to as Wet-Bulb depression) is small. If not, investigate for cause.
b. *Multiple Machine Operation* — Check to see if more plant capacity is in use than required for the conditions existing at the time.
c. *Outside Air Cycle* — Notice if excess ton-hours are experienced when outside temperatures are below 65° F. It is possible that full advantage is not being taken of the "outside or ventilation air-cycle" provided by the air handlers, particularly durin the night in those localities where high humidity does not normally prevail at night. The intelligent use of this cycle can result in appreciable dollar savings.
d. *Responsibility* — Supervisory personnel responsible for the operation of the air-conditioning system should make certain they read and become familiar with the subject of Air Handlers covered in Chapter 6. This is necessary in understanding the operating cycles of air handlers and in placing in effect the principles of load control.

Cooling Targets

The "ton-hours" referred to here are nothing more than the time-in-hours a refrigeration machine is operating, multiplied by the average tons of refrigeration it is producing during this time.

There are two methods of establishing this base target:

Estimate — For new buildings or where it is desired to start immediately with the first day the refrigeration plant is cut-in, the base targets for a start may be obtained from the Target Ton-hours (Estimate) Chart. (See Exhibit 12-D.) The ton-hours obtained from this chart are inserted in column (3) of the Heating-Cooling Chart opposite each corresponding mean temperature shown in column (4). It must be remembered that this chart is only an approximation. If need for adjustment is found, see "Target Adjustment" explained following the temperature ton-hour grouping.

Actual — This method utilizes data from the "Refrigeration Plant Operating Log" (Exhibit 12-B). To obtain the targets in this manner, the refrigeration plant must have been in operation at least 30 days in order to obtain a representative range of the daily average Dry-Bulb temperatures and corresponding ton-hours. First, the total ton-hours from column (19) of the log form for the 30 days is grouped by the average 24 hour daily Dry-Bulb temperature (bottom of column 7). A sample grouping is shown below:

TEMPERATURE (Fahrenheit)—TON-HOUR GROUPING

	65°–69°		70°–74°		75°–79°		80°–84°		85°–89°		90°–94°	
	Temp., °F.	Ton-Hrs.	Temp., °F.	Ton-Hrs.	Temp., °F.	Ton-Hrs.	Temp., °F.	Ton-Hrs.	Temp., °F.	Ton-Hrs.	Temp., °F.	Ton-Hrs.
	66	1800	72	2820	79	4400	83	5540	86	6020	91	6860
	67	1980	73	3000	76	3680	81	4880	88	6420	93	7160
			70	2440	77	3920	80	4600	86	6020	94	7260
			74	3200	78	4140	84	5600	89	6600		
					76	3680	81	4880	85	5860		
					77	3920	80	4600	88	6420		
							83	5540	85	5860		
							84	5600				
Averages:	66.5	1890	72.2	2865	77.2	3960	82.0	5155	86.7	6170	92.7	7090

When the groupings have been completed, they are then averaged as shown below the sample grouping. The averages then become the "key points" for plotting base-target curve. (See Exhibit 12-E.) Ordinary graph paper is used for the average Dry-Bulb temperature and the left side, or vertical scale, is used for the average ton-hours. The ton-hour scale is governed by the ton-hours in the 90° to 94° F temperature group. Since the ton-hours in the sample grouping are less than 10,000, a scale of 1,000 to 10,000 was used. If the ton-hours were more than 10,000 but less than 20,000, a scale of 2,000 to 20,000 would have been used, etc. When the key points have been plotted on the graph, they are then connected lightly using a pencil. The line is then evened out to form a smooth curve. When thus completed, the ton-hours for all intermediate temperatures may be taken directly from the curve (as illustrated) and inserted in column (3) of the Heating-Cooling Chart, opposite its corresponding temperature shown in column (4).

Target Adjustment — If, after a full month's operation, the targets are found to be too far out-of-line initially (or at a later date, due to improved operation) the base target curve may be adjusted upward or downward to reflect the change. This is accomplished by increasing or decreasing the "key ton-hours" for the corresponding "key temperatures" by the required percentage increase or decrease and then re-plotting the base target curve. (Exhibit 12-E.)

Potential Savings

Exhibit 12-F shows the savings achieved by one government agency with the introduction of controlled-heating plant operation.

A "heating record" form was introduced in the fall of 1937 resulting in a reduction of 20 per cent

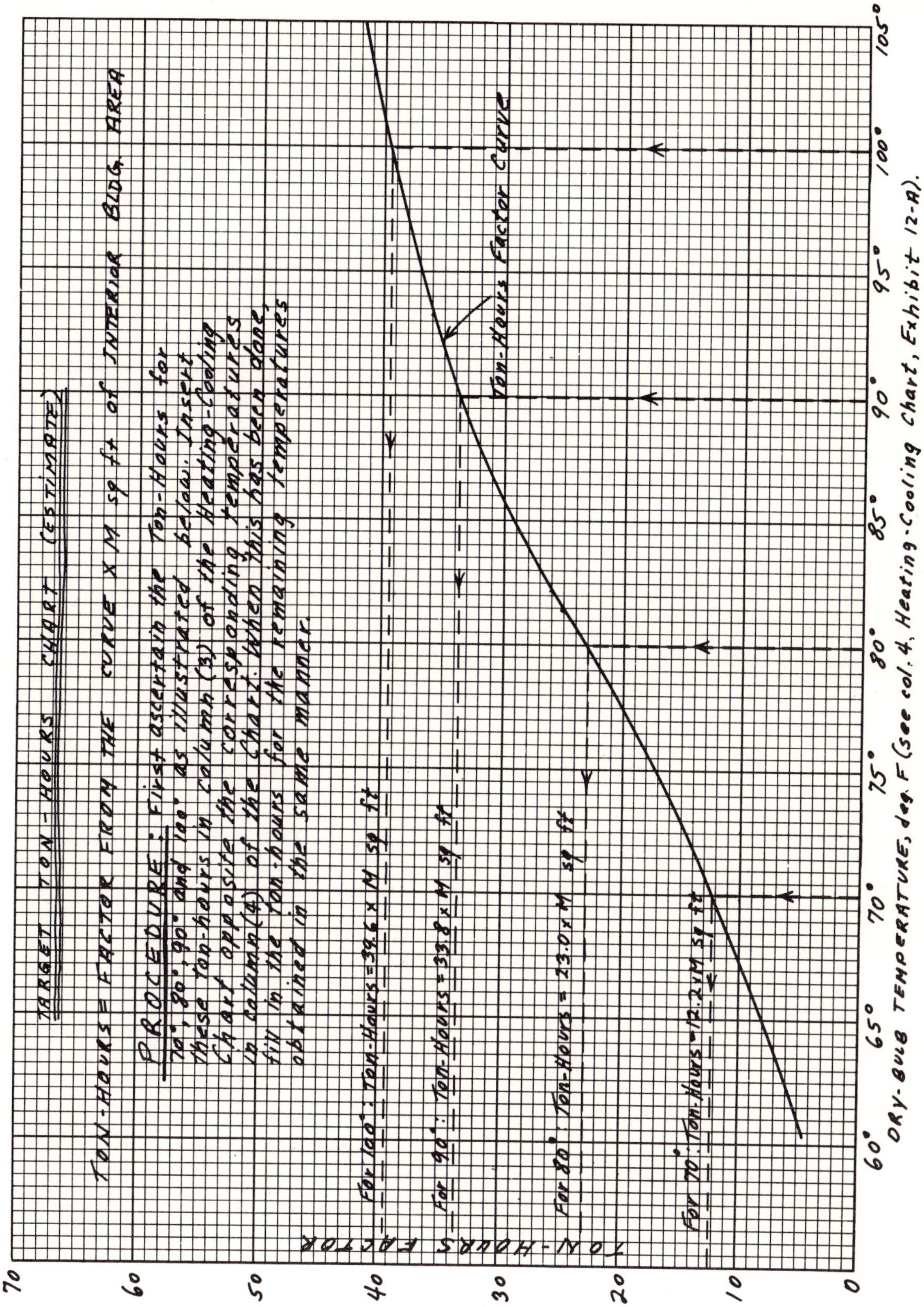

Exhibit 12-D. Target ton-hours chart (estimate).

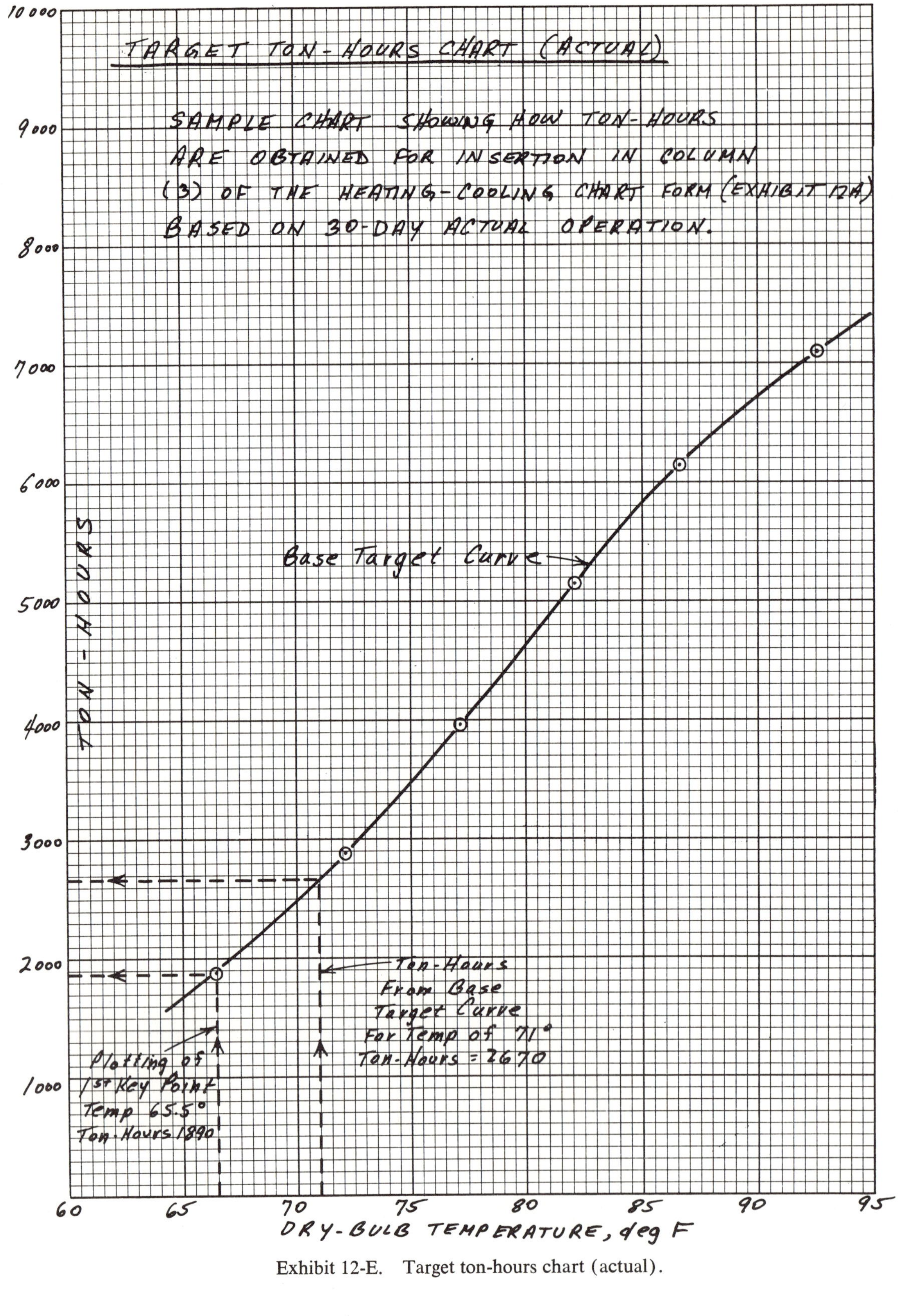

Exhibit 12-E. Target ton-hours chart (actual).

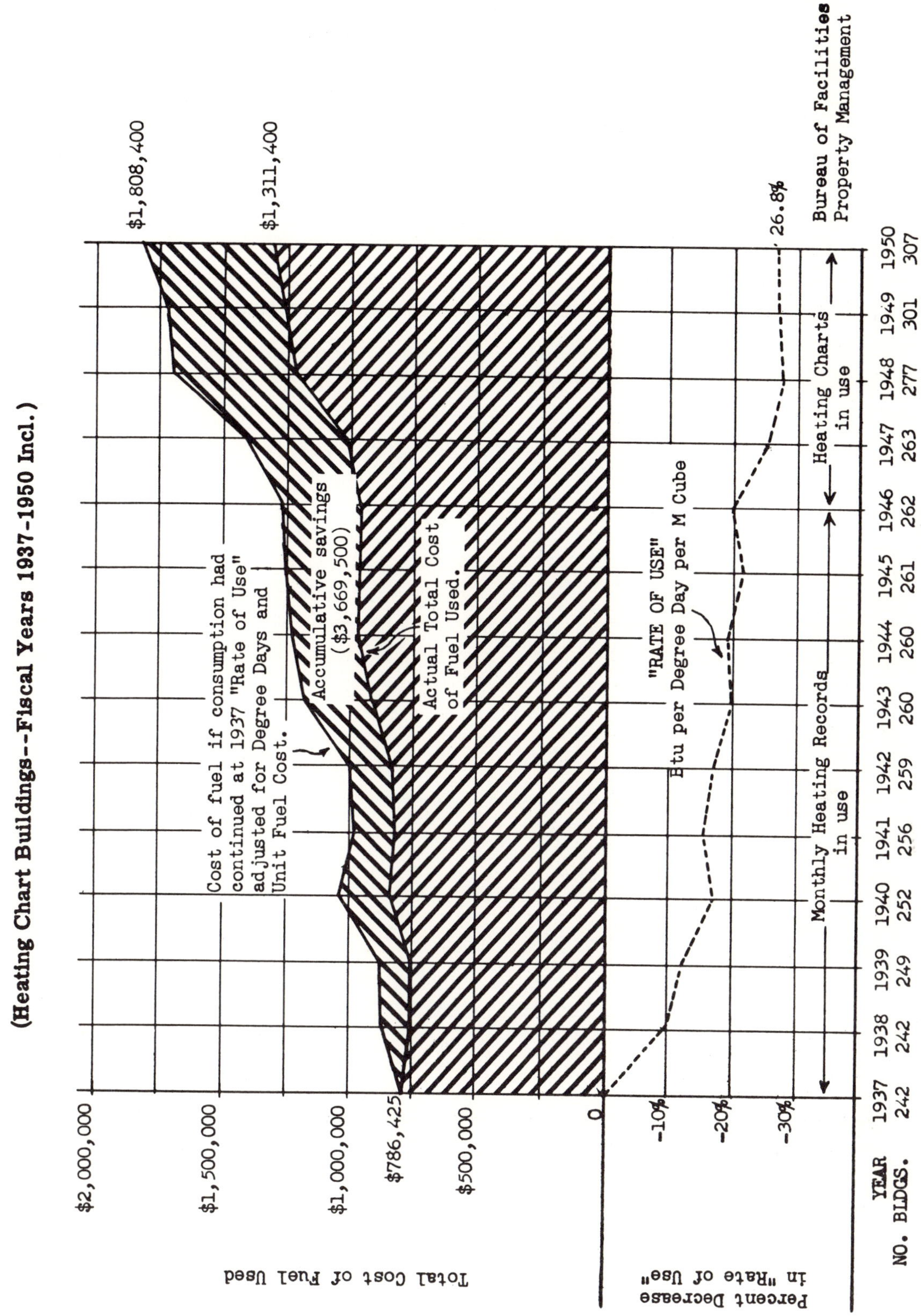

Exhibit 12-F. Dollar savings record. This chart not maintained subsequent to fiscal year 1950.

in the "rate of use" by 1943. After three years of level operation, a "heating chart" form was introduced whereupon a further reduction of 6.8 per cent was achieved. Note that these savings are for fuel costs only. Thus, it can be seen that, with comparable control over *both* the heating and cooling plants (and considering present higher fuel costs) the potential dollar savings through controlled plant operation as outlined herein, can be very rewarding.

Reference Material

The reference material as set forth below is comprised of both textbooks and maunfacturer's literature. These have been selected from among many as being of special value to the average building maintenance employee.

Before acquiring additional reference material, make certain there is on hand a copy of the manufacturer's "Operating and Maintenance Manual" for each piece of operating equipment, together with a list of essential parts. This type publication is the single, best source of operating and maintenance know-how for a particular piece of equipment.

Before requesting any of the material listed, inquiry should be addressed to the company as to its availability and cost, if any. The material is coded as follows:

B = Book
BL = Booklet
CL = Company Literature

American Society of Heating, Refrigeration and Air Conditioning Engineers, *Official Guide, Vol. II — Applications,* New York (B)

Association of Home Appliances Manufacturers, *Directory of Certified Room Air Conditioners,* Chicago, Ill. (BL)

Cooling Water on the Job; and *Water Treatment Service Tips,* Pittsburgh, Pa.: Calgon Corporation (CL)

Fundamentals of Refrigeration, Series of booklets with accompanying slides, if desired. Presents the complete picture of refrigeration machines — reciprocating type. Syracuse, New York: Carrier Air Conditioning Co. (BL)

"How to Conserve Water," Service Bulletin SB-105, Cleveland, Ohio: North American Mogul Products Company (CL)

Indoor Comfort System, Marshalltown, Iowa: Lennox Industries Inc. (BL)

Information for Expansion Valve, St. Louis, Mo.: Alco Valve Co. (CL)

Information Relative to the Expansion Valve, St. Louis, Mo.: Sporlan Valve Co. (CL)

Mechanical Refrigeration Equipment, Chicago, Ill.: American Oil Co. (BL)

Packaged Cooling Tower Know-how, Kansas City, Mo.: Marley Company (BL)

Power Factor Correction, North Adams, Mass.: Sprague Electric Co. (BL)

Reciprocating Refrigeration Manual; and *Air Conditioning Manual,* LaCrosse, Wisc.: The Trane Company (B)

Refrigeration and Air Conditioning Guide, New York: Theo Audel and Company (B)

Service Pointers, Wilmington, Del.: E. I. DuPont de Nemours & Co., Freon Products Div. (BL)

"Terminal Controls for Hydronic Air Conditioning Units," Detroit, Mich.: American Standard Corp., Controls Div., Covers 2-, 3-, and 4-pipe room fan-coil units. (CL)

The Electronic Handbook, Rockford, Ill.: Barber-Coleman Company (BL)

Index